## Curbing Catastrophe

What does Japan's 2011 nuclear accident have in common with the 2005 flooding of New Orleans from Hurricane Katrina? What does global warming have to do with "Black Hawk Down" Mark Bowden's classic account of a military disaster in Somalia? What is the relation between coal and tuna? This thought-provoking book presents a compelling account of recent and historical disasters, both natural and human-caused, drawing out common themes and providing a holistic understanding of hazards, disasters, and mitigation for anyone interested in this important and topical subject. Based on his on-the-ground experience with several recent major disasters, Timothy Dixon explores the complex science, politics, and economics behind a variety of disasters and environmental issues, and argues that many of the worst effects are avoidable. He describes examples of planning and safety failures, provides forecasts of future disasters, and proposes specific solutions for hazard mitigation. This book shows how billions of dollars and countless lives could be saved by adopting longer-term thinking for infrastructure planning and building, and argues that better communication is vital in reducing global risks and preventing future catastrophes.

TIMOTHY H. DIXON is a professor in the School of Geosciences and Director of the Natural Hazards Network at the University of South Florida in Tampa. In his research, he uses satellite geodesy and remote sensing data to study earthquakes and volcano deformation, coastal subsidence, ground water extraction, and glacier motion. He has worked as a commercial pilot and scientific diver; has conducted research at NASA's Jet Propulsion Laboratory in Pasadena, California; and was a professor at the University of Miami, where he co-founded the Center for Southeastern Tropical Advanced Remote Sensing (CSTARS). Dixon was a Distinguished Lecturer for the American Association of Petroleum Geologists (AAPG) in 2006–2007. He is also a Fellow of the American Geophysical Union (AGU), the Geological Society of America (GSA), and the American Association for the Advancement of Science (AAAS). He received a GSA Best Paper Award in 2006, and received GSA's George P. Woollard Award in 2010 for excellence in geophysics.

# Curbing Catastrophe

## Natural Hazards and Risk Reduction in the Modern World

TIMOTHY H. DIXON
University of South Florida

**CAMBRIDGE**
**UNIVERSITY PRESS**

University Printing House, Cambridge CB2 8BS, United Kingdom

One Liberty Plaza, 20th Floor, New York, NY 10006, USA

477 Williamstown Road, Port Melbourne, VIC 3207, Australia

4843/24, 2nd Floor, Ansari Road, Daryaganj, Delhi – 110002, India

79 Anson Road, #06–04/06, Singapore 079906

Cambridge University Press is part of the University of Cambridge.

It furthers the University's mission by disseminating knowledge in the pursuit of education, learning, and research at the highest international levels of excellence.

www.cambridge.org
Information on this title: www.cambridge.org/9781107035188
10.1017/9781139547345



First published 2017

Printed in the United Kingdom by TJ International Ltd. Padstow Cornwall

*A catalogue record for this publication is available from the British Library.*

ISBN 978-1-107-03518-8 Hardback

Additional resources for this publication at www.cambridge.org/dixon

*For my grandchildren and their grandchildren*

# Contents

# Online Appendix

Available at www.cambridge.org/dixon

# Preface

*"Civilization exists by geological consent, subject to change without notice."*

Will Durant, American philosopher and writer (1885–1981)

We have a troubled relationship with our home planet. While we depend on it for the air we breathe, the water we drink, the soil that grows our crops, and the minerals and energy we consume, we also exploit and despoil its resources with little thought for the future. We're rapidly depleting freshwater aquifers and mineral deposits that took the Earth thousands to millions of years to develop. Too often, we build our homes, factories, and power plants in vulnerable areas, next to earthquake faults, volcanoes, and low-lying areas prone to flooding. We use energy in ways and amounts that have harmful consequences. Rather than choosing to fit ourselves into the planet's ecosystem and use its resources sparingly, we have chosen a more confrontational approach.

Astronomical observations suggest that numerous planets in our galaxy have the conditions necessary for life. Given the immense size and age of our galaxy, we might expect that at least a few of these planets have developed advanced civilizations, but there is no evidence of this. The conundrum is known as Fermi's Paradox, after atomic scientist Enrico Fermi, who famously said, "Where is everybody?" Astronomer Carl Sagan speculated that any planet that developed an advanced civilization would soon destroy itself, either through war or environmental degradation.

Politicians and diplomats have the job of preventing war. Scientists have the job of educating the public and government officials about the planet and its ecosystems, and how we can do a better job of "fitting in" to our environment. Designing, locating, and building our infrastructure so that it's tough enough to meet nature's demands, but is also kind to the planet's ecosystems, are important aspects of fitting in. This book describes some of our most important successes and failures in these areas.

In the last decade, three major disasters unfolded in areas where I have some professional background: the 2005 flooding of New Orleans associated with Hurricane Katrina; the 2010 earthquake in Haiti; and the 2011 earthquake and tsunami in Japan. I had worked in New Orleans and Haiti prior to those two events, doing research on the geological processes that ultimately contributed to the disasters (subsidence and sea level rise in New Orleans; plate motion and earthquake strain accumulation in Haiti). In the two cases I worked on directly, I was struck by the fact that scientists "in the know" knew these disasters were inevitable, but felt powerless to do anything about them. The public was largely ignorant, and government officials, policy makers, and in Haiti's case, aid agencies, were focused on issues deemed more immediate. Many businesses were uninterested or even hostile to the warnings of scientists. Similar problems appear to apply to Japan's recent nuclear disaster and to many environmental issues. A big part of the problem is communication, or more specifically, its lack. Many scientists are not very good at communicating their knowledge of risk to the public, including the long-term consequences of building things in harm's way or designing them to inadequate standards. Equally culpable, some politicians, some members of the media, and some members of the public are not especially good at listening to scientists who discuss these issues. Scientific literacy seems to be on the decline in modern society, which is odd given our increasing use of and dependence on technology that ultimately comes from science. Aspects of the current debate about energy use and global warming similarly reflect poor communication among scientists, policy makers, the media, and the public, and failure of the last three groups to adequately consider long-term trends.

This book attempts to address these problems. It discusses issues on the boundary between science, business, and government, including costs, risks, and mitigation strategies for natural and human-made hazards. I describe examples where scientists have been called on to give policy advice on these hazards, why this advice is sometimes taken and sometimes ignored, and the economic and human consequences of following or ignoring that advice. I describe examples where scientists got things right and where they got things wrong. I suggest some reasons why humans tend to ignore obvious, long-term problems; suggest how scientists might overcome this tendency; and make specific suggestions on how to improve information flow, decision making, and transparency. In addition to hind-casts (in-depth forensic analysis of past disasters, known to my critics as Monday morning quarterbacking), I also make some forecasts. Some of these forecasts will be controversial, but hopefully will spur debate, additional research, and much-needed public and private action.

This book has three main themes, all related to natural or human-made disasters: the importance of communication, the importance of understanding long-term processes and time scales, and the economic consequence of failure. I'll wrestle with what long-term means in later chapters, but for now let's define it as anything longer than a typical human life span. The concept of time lag is another recurring theme. It's used to explain why cities are often located in areas now considered unsafe (Chapter 3), why the price of oil goes through such large swings (Chapter 6), and why it's so difficult for scientists to explain the dangers of global warming (Chapter 8).

Like most authors, I have some biases. I am a geologist, so I tend to view many issues through the prism of Earth science and the common sense approach inherent to my discipline. Although the book aims at a global focus, I live and work in the US, a country that – because of its size, location, and geologic setting – experiences many natural hazards, including hurricanes, tornadoes, earthquakes, tsunamis, and volcanic eruptions. This range of extreme natural phenomena makes the country a useful "natural laboratory" for some of our discussions.

This book doesn't focus exclusively on natural hazards, nor does it provide a catalog of disasters. Rather, I focus on a few examples to illuminate common themes. Natural disasters, as well as the human-made kind, often share several features, including poorly designed infrastructure. Shining a spotlight on these common features can curb catastrophes for future generations. Otto von Bismarck, chancellor of Prussia and later Germany in the late 19th century, once said, "Only a fool learns from his own mistakes. The wise man learns from the mistakes of others." My hope is that this book will help all of us learn from past mistakes.

The book is intended for a general audience. High school math – or better yet, simple common sense – is adequate preparation. Throughout the book, separate sections ("Boxes") provide further detail for interested readers; consider it my attempt at a "Director's Cut." An online appendix entitled "Exercises for Students" is available as an electronic supplement from the publisher. It includes additional background material and example problems that are suitable for introductory undergraduate classes in natural hazards, geology, geography, climate change, and environmental science.

Chapter 1 gives an introduction and overview of major issues. Some of these are covered in more detail in later chapters, such as the cost of the earthquakes in Japan and Haiti, and the various missteps that contributed to these horrific disasters. Chapter 2 discusses what I mean by the term "natural hazard" and describes the underlying science. Readers with a background in Earth science will be familiar with most of this material. This chapter also makes the case that natural and human-made disasters are not so different. Both often have root causes in things like bad design, bad engineering, bad management, or failure to think long term. The solutions to preventing both types of disasters are also often similar (improved infrastructure). Chapter 3 looks at why our scientific understanding of hazards does not always translate into effective action. I emphasize the importance of having a good long-term record when assessing a given hazard, and I provide an introduction to the measurement techniques that scientists use to obtain these records. Chapter 4 applies the lessons of Chapters 2 and 3 to the recent Japanese earthquake, tsunami, and nuclear plant failure, and describes an eerily similar event that took place years earlier that should have been a wake-up call for the nuclear industry. I argue that both nuclear disasters were preventable given knowledge available at the time, a position that remains controversial. Chapter 5 describes the possibility of future earthquake and tsunami disasters in several other areas, including the US Pacific Northwest; Geneva, Switzerland; and Istanbul, Turkey.

Chapter 6 considers the larger issue of nuclear power relative to other power sources, including coal, and uses this issue to introduce the concept of relative risk. I argue that nuclear power should remain on the energy table as an important carbon-free option, a stance that is not popular with many environmentalists. Chapter 7 discusses flood disasters, including examples from the US (Galveston, Texas and New Orleans, Louisiana) and Bangladesh, and considers how future flood risk will be exacerbated by sea-level rise. Chapter 8 reviews the cause of sea-level rise (global warming) and how this topic got to be so contentious. This chapter makes the case that global warming has similarities to some of the other more immediate disasters discussed in earlier chapters by focusing on the themes of scientific communication and the importance of considering long-term processes and economic consequences. Improved infrastructure (in this case for transportation and energy generation) is also relevant. The last chapter offers specific solutions, including a summary of recommendations.

Communication is a major theme of this book and is mentioned in almost every chapter. It is a two-way street: It is important for scientists and engineers to clarify complex subjects and crucial for the media, the public, and government officials to be active and engaged listeners. Communication also means different things to different people, and there are different levels of communication, from being informed to being strongly engaged. At the most basic level, it means transferring useful information. In several places, I assert that scientists need to do a better job of transferring information into the public domain. As an example of a deeper level of communication, a citizen activist could take information from this book, such as the scientific and engineering background related to the Japanese earthquake and nuclear disaster, or the environmental consequences associated with coal-fired power plants, and use it to influence public policy.

## NOTES AND REFERENCES

Most nonfiction books and scientific articles make extensive use of footnotes and references that list relevant published work. While this allows interested readers to follow up on specific issues and gives proper credit to previous researchers, it can also make for cumbersome reading.

Here I've avoided footnotes and most references within the text wherever possible. Instead, relevant work for each chapter is compiled at the end of the book. In most cases, the link between specific points in the text and a given author's work should be clear from the title. In cases where this is not clear, I have either called out specific references in the text, with the author's last name and publication year in parentheses, or made a comment in the reference list itself.

Commented references are indicated by an asterisk (*).

## COLOR FIGURES

In order to keep the book more affordable, figures are reproduced in black and white. The original versions of some of these figures are in color. The color versions are available on the author's web site (http://labs.cas.usf.edu/geodesy/)

## CONFLICTS OF INTEREST

Throughout the book, and especially in the last chapter (Solutions), I make a case for transparency, particularly with regard to funding sources. If people have access to more information and are aware of potential conflicts of interest when experts give their opinion, then better, more informed decisions can be made. So here goes for me: In the last two decades, I have received research funding from the US National Science Foundation (NSF), the National Aeronautics and Space Administration (NASA), the Office of Naval Research (ONR), the Department of Energy (DOE), and British Petroleum (BP). I have also received an honorarium from the American Association of Petroleum Geologists (AAPG) for a series of public lectures. I have been involved in the governance of two scientific organizations, the American Geophysical Union (AGU) and the American Association for the Advancement of Science (AAAS), and am also a member and Fellow of the Geological Society of America. Where specific conflicts exist, I will call them out in the text, usually beginning with the phrase "full disclosure."

# Acknowledgments

Many people contributed ideas and time to this book, and I'd like to thank some of them here.

My wife, Dr. Jackie Dixon, a professional scientist with her own career and responsibilities, spent many hours reviewing early drafts and saved me from saying some really stupid things.

Dr. Robert Stern, a friend, colleague, and Professor of Geological Sciences at the University of Texas at Dallas, suggested I investigate Alaska's Nenana Ice Classic.

Dr. Grace Palladino had the misfortune of wading through an early draft, but nevertheless provided numerous suggestions and many encouraging words.

My friend and neighbor Rick Ker also reviewed an early draft and prevented a large number of typographical errors and mangled phrases.

Dr. Giovanni Sella re-drafted Figure 2.2 from its original color version to make it legible in black and white.

Dr. Louise Ander re-drafted Figure 6.2 from its original color version to make it legible in black and white.

John Hofmeister, former President of Shell Oil USA and author of *Why We Hate the Oil Companies*, provided a review and advice on how to avoid legal issues. Time will tell how well I followed his advice.

I am indebted to Dr. Charles Connor who brought to my attention the disturbing case of the Kashiwazaki Kariwa nuclear accident that occurred several years before the events at the Fukushima Daiichi nuclear plant.

Drs. John Adams and Chris Goldfinger provided critical background on the Cascadia Margin story and some great quotes.

Dr. Richard Somerville, a climate scientist at Scripps Institution of Oceanography, and Mark Bowden, author of *Black Hawk Down*, gave technical reviews of parts of the book, for which I am extremely grateful.

Dr. Amelia Shevenell gave me a number of important comments and an excellent figure illustrating long-term climate cycles.

I am indebted to Dr. David Chapman for suggesting one of the climate problems, and to Dr. Geoffrey Vallis for his delightful "primer" on oceans and climate, and for two other problems in the climate section (any errors are strictly my own).

Dr. Robert Yeats and two anonymous reviewers provided comprehensive reviews that greatly improved the book.

Dr. Emma Kiddle was my editor at Cambridge University Press and made a large number of useful suggestions, most of which I've followed.

Susan Santucci, founder of Ivy League Tutoring, kindly provided a final read and clean-up of my sometimes tortured prose.

Many figures in the book were prepared or improved by Theresa Maye, Design Magic Studio.

Parts of this book summarize my own research, which has been generously supported by the US National Science Foundation (NSF), the National Aeronautics and Space Administration (NASA), the Department of Energy (DOE), the Office of Naval Research (ONR), and internal funding from the University of South Florida (USF).

While parts of this book have been critical of the media, I recognize that there are some real heroes out there. Many journalists work hard to get the message out in the face of tight budgets, impossible deadlines, withering criticism, and even threats from special interests or thuggish governments.

# 1 Black and White Swans, Communication, Evolution, and Markets

*"What we've got here is failure to communicate."*

Dialogue from the 1967 American film "Cool Hand Luke"

Japan's nuclear disaster of 2011 caused huge economic losses, not only for Japan, but for the world economy. While the earthquake and accompanying tsunami that caused problems at the Fukushima-Daiichi nuclear complex were unavoidable, the accompanying nuclear and financial meltdowns were not. Many geologists were aware of the possibility of large tsunamis in this part of coastal Japan, but engineers and government regulators did not consider the possibility of such high water or factor it into the plant's design. Backup power systems failed in the accompanying flood, leading to the overheating of fuel rods and a nuclear plant operator's worst nightmare, a core meltdown. The miscommunication between scientists and plant managers contributed significantly to the problems at the power plant, the resulting nationwide power interruptions, and the consequent factory closings and supply chain disruptions around the world. Although it is difficult to separate total losses from the earthquake and those due only to the nuclear plant, it is likely that the latter caused Japanese and global losses exceeding US $100 billion (all dollar amounts are in US currency unless otherwise indicated). In an increasingly complex and interdependent world economy, we can no longer afford such easily avoidable mistakes.

In 2007, Nassim Taleb published an influential book called *The Black Swan*. Taleb's basic premise is that in many human endeavors, we seek to manage risk by preparing for a range of possible bad events. These hypothetical "worst case" scenarios may be based on past experience, or expert predictions. While this approach works most of the time, every once in a while things go spectacularly wrong; the worst case event envisioned by planners was not nearly bad enough. Such events are sufficiently rare that they are difficult or impossible to predict, hence difficult or impossible to prepare for. Taleb refers to such events as "black swans," since most swans are white, and we come to expect the latter color. Many recent catastrophes – from the financial meltdown in 2008–2009 to Japan's earthquake, tsunami, and nuclear meltdown in 2011 – have been called black swans.

While some disasters may exemplify Taleb's black swans, I argue that many, at least the natural ones, should have been foreseen. While we can't stop an earthquake or a hurricane, we ought to at least be able to mitigate their worst effects through basic planning and better building. Earthquakes, tsunamis, and hurricanes usually occur in well-defined places for well-understood reasons. Some of the problems in New Orleans in 2005 and at Japan's Fukushima nuclear plant in 2011 have strikingly similar causes. Critical facilities in New Orleans lost power because backup generators or their control functions were in the basement or ground floor, easily knocked out by floodwaters. Similarly, backup power and certain control functions at the Fukushima power plant were sited at low elevations, susceptible to flooding. In both cases, many experts knew that flooding at these locations was possible.

If most swans are indeed white, we should be able to plan accordingly. Why we don't has a number of explanations, but two important ones relate to communication. First, while scientists have a pretty good understanding of the key processes that produce natural disasters, we are not always very good at communicating that understanding to the public and to policy makers. Our writing and speaking styles can be long-winded and verbose. We often violate Rule # 1 of *Elements of Style*, Strunk and White's classic treatise on clear writing. (Rule # 1 is "Use fewer words.") Second, government and policy experts, business, and the media are not particularly good at listening to scientists. As a result, scientific understanding of risk often fails to translate into effective government, business, or personal preparation. There are too many barriers between scientific knowledge and appropriate action.

One reason that it's hard to get the attention of the public and policy makers is that scientific understanding of process does not necessarily translate into predictive capability. It's tough to get everyone's attention if you can't come up with a clear "do something by this date" statement. Scientists could not predict the timing of the 2011 earthquake and tsunami in Japan or the strikingly similar events in 2004 in Sumatra. Scientists knew that large earthquakes and tsunamis were possible at these locations, but they could not say when they would occur. Given human nature, it is easier to act on threats perceived as immediate; long-term threats are often ignored. Being told that a major earthquake is likely within the next 50 years gives policy makers sufficient time to change building codes and the construction industry time to strengthen buildings, but often that knowledge does not translate into action. Political cycles are short, and next year's budget problems usually trump longer-term considerations. The same considerations often apply in business. The tenure of company CEOs has been getting shorter as investors demand quick profits. Better to leave those boring but costly infrastructure upgrades to one's successor.

In terms of the public's ability to listen, many natural phenomena (my preferred term for natural disasters) occur on time scales that are long when compared to typical human experience; they are off the radar screen for most people, even many scientists. A colleague trained in astronomy expressed surprise when I explained to him that he was involved in a project to build an astronomical observatory on a mountain in Mexico that was part of an active volcano. Although it had not erupted in several hundred years, there was no guarantee that it would not erupt again within the next 50 years, the lifetime of the instrument being designed.

To some extent, this lack of appreciation for long time scales reflects our evolutionary heritage. This means that two themes of this book are closely related: The less than superb communication skills of many scientists, combined with the innate human characteristic of focusing on short-term issues, means that non-scientists tend not to listen to our message, especially when that message involves long-term trends and long-term risks.

## LONG-TERM RISK, LISTENING SKILLS, AND EVOLUTION

Most people know that it is not healthy to smoke too many cigarettes or eat too much junk food (foods high in sugar, salt, or fat). Too much smoking leads to lung cancer, while eating too much junk food can lead to obesity, diabetes, stroke, and heart attack. Both practices can lead to premature death. Yet businesses make huge profits every year selling cigarettes, junk food, and sugary drinks. They know that people can be relied upon to choose short-term pleasure despite long-term risk. We have a built-in "present bias."

Why do people discount or ignore long-term risk? There are a number of explanations, but I'm going to focus on one that I think is important, at least when applied to discussion and understanding of longer time scale Earth processes: human nature, bequeathed to us by millions of years of evolution. For most species, including our ancestors, survival depended on sensitivity to immediate surroundings and events. Even today, most of us are hardwired to be sensitive to things happening around us; we are not hardwired to be sensitive to change on longer time scales. As with any inherited trait, this reflects two sorts of selection pressures that influence evolution: survival selection and sexual selection. In terms of survival, when our ancestors wandered in the wilderness, hunting and gathering, they had to be acutely aware of their immediate surroundings and *changes* to their surroundings – survival depended on it. But not changes on time scales of decades or centuries. Changes on time scales of days or seconds were the things that mattered. Even when our ancestors turned to farming several thousand years ago, they mainly had to be sensitive to seasonal changes; longer-term processes were simply not important when life spans were shorter than 30 or 40 years. For most species, there is little survival advantage bestowed upon individuals who are sensitive to changes on decade or longer time scales – it just doesn't matter. To this day, a geologist talking about time scales of hundreds or millions of years to a room full of non-geologists is sure to be met with glazed eyeballs. In terms of sexual selection, let's be honest: The successful hunter of 100,000 years ago had a better chance of getting the girl than the guy who waxed eloquent about a volcanic eruption flattening the village years in the future.

We all have a tendency to focus on disasters – they are immediate, and more interesting. The media and the audience are unlikely to focus on examples of earthquakes and hurricanes where planning and well-designed buildings did their job of minimizing casualties and economic disruption; such events are not considered newsworthy. "If it bleeds, it leads" is the newsroom rule. The low- to no-impact examples are, nevertheless, important for us to consider, demonstrating that we can build smarter. Moderate to large earthquakes happen frequently in Japan, New Zealand, California, and Chile, and the well-designed and well-built infrastructure in these locations usually performs quite well. When things do go wrong, we can study the failure and improve things, whether it's the basic design, construction practice, or management. This is essentially the focus of this book. Many studies suggest that relatively modest changes in design and construction can improve survivability of infrastructure. I'll try to identify this "low-hanging fruit" in subsequent chapters.

On September 5, 2012, a magnitude 7.6 earthquake struck northwestern Costa Rica. Although this was a large earthquake, damage was not significant, and there were no fatalities (one person had a heart attack near the time of the earthquake, which may or may not have been earthquake-related). Part of the explanation for the lack of damage and minimal loss of life was that this earthquake was expected, at least in a general sense. Beginning in the 1990s, Costa

Rican seismologist Dr. Marino Protti began to publicize the fact that, given the past history of large earthquakes in this area, the probability of a similar event in the future was high. Although he did not predict an earthquake in any given year, he worked tirelessly to warn the public and government officials about the likelihood of large earthquakes and the importance of adequate building codes and preparation. The activist approach by Protti and his colleagues to earthquake preparedness had a significant impact on disaster mitigation in the region and is an indicator that it is possible to do better, even with limited resources.

Prior to catastrophe, it is difficult to get the attention of the media on natural hazards in the crush of other more immediate events. Science in general, and natural hazard planning in particular, are perceived to be less important than military or national security issues, not as exciting as the next sports match, not as sexy as the latest celebrity scandal. A geologist using public funds to dig trenches is more likely to be fodder for late night comics than a media darling, even if that trench contains clues to the last earthquake or tsunami, information that could save many lives and billions of dollars in avoided damage.

### BOX I.I Digging trenches and studying dirt: The principle of superposition

How do geologists get information on past disasters if the events are too old to be described in history books? It turns out that nature leaves clues in the top few meters (10 to 15 feet) of the Earth's surface. By doing some detective work and studying surficial soil and sediments (the "dirt"), we can often piece together a sequence of events from the clues left behind. For example, geologists studying past tsunamis may dig a trench to look in cross section at layers of sediment that were deposited by the tsunami. The basic principle is "superposition," first enunciated by Danish scientist Nicolas Steno in the 17th century. This simple idea states that sediments on the

BOX 1.1 **(cont.)**

top are usually younger than sediments buried more deeply. By dating material lodged within a given disturbed layer with radio-carbon or other technique, it is possible to estimate the age of the last tsunami and, if several events are visible, get an idea of the frequency of these events. This is an incredibly valuable tool for hazard assessment and is one reason why geologists are so enamored of long time scales and digging trenches, as described in the next few chapters and Figure 4.1

## COST OF DISASTERS AND THE ROLE OF MARKETS

If communication is part of our problem concerning natural disasters, how important is it to fix it? Experience shows it is hugely important. Consider losses associated with Japan's 2011 nuclear disaster. Total costs from the earthquake and tsunami will probably exceed $200 billion. Some of these costs were unavoidable, but costs associated with the nuclear accident (the avoidable part) will probably exceed $100 billion. These include the direct costs associated with cleanup, lost power and its replacement cost, and lost production in industries reliant on that power, as well as indirect costs related to future lawsuits or payments to individuals and businesses damaged in the disaster, lost economic opportunity associated with abandonment of the irradiated zone, lost economic opportunity by industries forced to close or reduce output after the disaster, and worldwide economic disruption. (In today's global economy, many factories around the world rely on products produced in Japan.) The accident affected the nuclear industry worldwide, making it more difficult to gain approval for new nuclear power plants, a blow to the reduced carbon energy strategy in some countries (Chapter 6). Germany decided to phase out nuclear power as a direct result of the Japanese accident. Natural disasters are expensive, not only in terms of injury and loss of life but also in damaged or destroyed infrastructure, lost business, and reduction in future economic growth. In a world experiencing increasing fiscal problems, we cannot afford to waste such large sums of money on preventable disasters.

Disasters are getting more expensive with time. Does this reflect increased disaster frequency (Figure 1.1), or are other factors at play? One factor contributing to the increased cost and frequency is population growth. In other words, even if disasters are occurring at the same rate, there are now more people in harm's way. Figure 1.2 shows US costs from natural disasters over the same period as Figure 1.1. Cost increased more than fivefold since 1980, during a period when US population increased by only about 35 percent. The difference reflects several factors, including the increasing percentage of population concentrated in vulnerable coastal areas

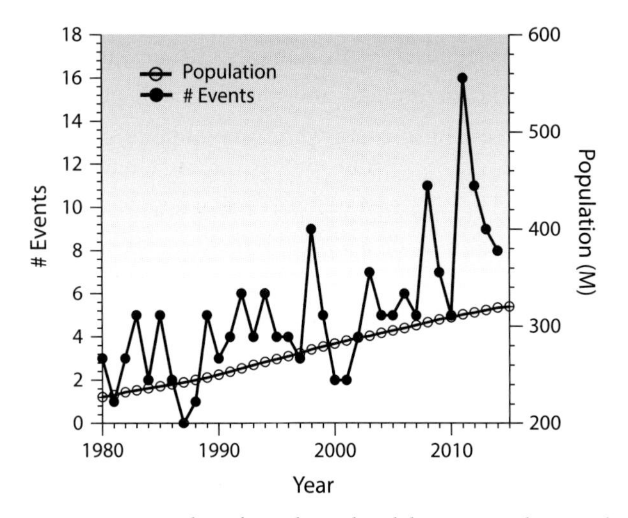


FIGURE 1.1 Number of weather-related disasters in the US whose cost exceeded $1 billion (left-hand scale), compared to the US population in millions (right-hand scale). The horizontal scale shows time for the years 1980 to 2014, the period when good data on disaster costs are available. Costs adjusted for inflation to 2015 dollars. Disaster data from National Atmosphere and Ocean Administration (http://www.ncdc.noaa.gov/billions/). Population data from World Bank.

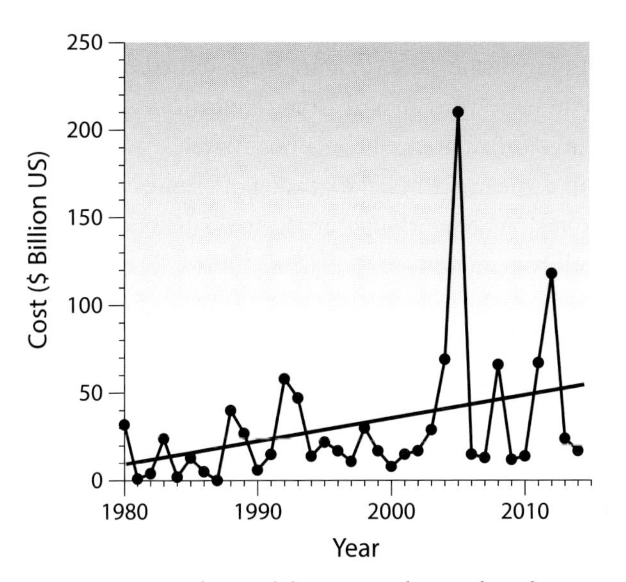


FIGURE 1.2 Cost of natural disasters in the US plotted against time (years from 1980 to 2104, the same period as Figure 1.1), adjusted to 2015 dollars. A best-fit line through the data is also shown. Data from NOAA.

that are susceptible to storm-induced flooding and tsunamis, often the costliest disasters. The value of coastal real estate and other coastal infrastructure has also increased significantly over this period. The intensity of storms may also be increasing (likely a consequence of global warming), which in turn increases the damage from a given storm (Chapters 7 and 8).

Different countries may have very different costs due to population growth trajectories (how fast their population is increasing) and their geography (for example, the amount of vulnerable coastline). The Philippines is especially susceptible to flooding from storm surge – it lies in the tropical typhoon belt and, being an island nation, has lots of coastline. It also has a rapidly growing population, with a large percentage living in vulnerable coastal areas.

Many low-lying coastal countries will be increasingly affected by sea-level rise and hurricane-related storm surge, so costs associated with these events are likely to continue rising. In the US, populations vulnerable to hurricanes and flooding include areas along the Gulf of Mexico and the Eastern Seaboard. In the tectonically active west coast of the US, large urban areas such as Los Angeles, San Francisco, and Seattle lie on or near earthquake-prone fault zones. Disasters in any one of these regions have the potential to deal a devastating blow to the US and global economy.

To illustrate the increasing cost with time, consider the recent US experience. In 2011, a record-breaking 16 disasters that exceeded the billion dollar cost threshold occurred, breaking the previous record (11 in 2008). NOAA estimates that the US experienced approximately $67 billion in damages from weather-related events in 2011 and nearly 1,000 fatalities from nine severe weather or tornado incidents, two flood events on the Mississippi and Missouri Rivers, one drought/heat wave, one winter blizzard, one wildfire, and two tropical storms and hurricanes, including Hurricane Irene. Munich Re, a global re-insurance firm that insures the insurance companies in the event of large losses, estimated 2011 costs in the US at $77 billion. This counts insured plus uninsured costs, and uses a slightly broader definition of disaster compared to the NOAA estimate. The year 2005 continues to hold the US record for total cost: approximately $200 billion according to NOAA. Damages from Hurricane Katrina alone exceeded $100 billion (Lott and Ross, 2006). Costs for 2012, totaling approximately $118 billion, include significant costs from a summer drought and heat wave (mainly agricultural losses and wildfires), and storm and flood damage from Hurricane Sandy in October. The biggest wildfire season on record occurred in 2015. Total costs since 1980 exceed $1 trillion.

If present trends continue, by 2050 the US can expect to incur an average annual cost of about $150 billion from weather-related disasters, with the occasional (possibly once per decade; the exact frequency cannot be known) loss of half a trillion dollars. To paraphrase

Everett Dirksen, a US senator famed for his opposition to wasteful spending, this is starting to sound like real money.

Many of these costs are avoidable. We inflict them on ourselves (or more accurately, on our children and grandchildren since the money to repair damage is partly borrowed) through bad planning and design, poor construction practices, poor management decisions, and poor political decisions. To borrow a term from the sports world, we might call them societal "own goals," which happens when a player scores against his or her own team. For our children and grandchildren's sake, we need to score fewer own goals.

Clearly, we need better ways of translating scientific knowledge of risk in vulnerable areas into a built environment that is safe, cost-effective, and resilient. And when we get it wrong, we need to rebuild smarter, learning from the past rather than making the same mistakes over and over. After catastrophic flooding in the Mississippi River Valley in 1993, some communities simply rebuilt damaged or destroyed infrastructure in the same flood-prone areas (there is a reason that geologists call these areas "flood plains"). Although it is difficult to make predictions, we can be certain that those communities will lose their rebuilding investment at some point in the future. Engineers who study flood control levees point out that there are two types: those that have failed and those that will fail. Fortunately, many other communities took the opportunity to convert flood-prone areas to green space or low-cost, easy-to-repair facilities such as parking lots. This is a trend that needs reinforcement.

At the present time, there are few market mechanisms to reward communities and businesses that plan and build smarter. In many Western countries, poor planning and design may actually subsidized by taxpayers. Local and state governments assume that national governments will bail them out when natural disaster strikes. In the US, declaring a state of emergency gives communities and individuals access to grants and subsidized loans. Until recently, private businesses could get low-cost flood insurance that was subsidized by the government. However, as the cost of natural disasters increases, such subsidies will be increasingly difficult for governments to bear. Eventually it will be recognized that long-term investments are required to avoid or mitigate disasters. An investment of about $50 million (2,000 times cheaper than the eventual cost) and better management practices would have greatly reduced or eliminated the risk of backup power loss at Fukushima and many subsequent problems. Eventually, private business will start to see such investments as sound business practice, and be proactive. Governments will start to withdraw the risk subsidy that promotes bad building and investment decisions, as the cost of these subsidies increases. Markets will begin to reward well-planned and safe communities, e.g. through lower insurance premiums and increased business investment. At the present time, you cannot buy shares in a mutual fund that only invests in factories, power plants and other businesses certified by independent scientists and engineers as safe; perhaps such funds will exist in the future. In principle, governments provide this assurance via zoning and other regulations, but unfortunately these regulations and their enforcement are subject to political influence. The insurance industry is increasingly pricing the risk of natural hazards into its premiums, using models that are increasingly accurate. Perhaps in the future, borrowing costs will be higher for cities, businesses, and individuals whose locations are deemed at high risk from natural disasters if they have failed to invest in adequate mitigation, e.g. New Orleans (from hurricanes and flooding, Chapter 7), Seattle (from earthquakes, Chapter 5), or low-lying parts of Geneva (from tsunamis, Chapter 5). In summary, there are large future financial costs for many preventable disasters, and past market failures have contributed to this. I suspect this partly reflects bad or incomplete information, a problem I hope this book can reduce.

The next two chapters address the issue of incomplete information and how it impacts risk assessment and disaster preparation. Chapter 2 discusses the background science for some common disasters (earthquakes, volcanoes, tsunamis, and hurricanes). Chapter 3 considers the following question: If we understand the background science, why are we so often struck by disasters?

# 2 What Is a Natural Disaster? Where Do They Occur and Why? Are They Different from Human-Made Disasters?

In 2015, a magnitude 5.9 earthquake struck the west coast of the island of Sabah in Malaysia, killing at least 16 people. The government of Malaysia blamed a small group of Canadian, Dutch, and German tourists who had visited nearby Mount Kinabalu and stripped naked to celebrate their ascent of the peak, taking photos of themselves and posting them on social media. The government felt that the tourists had disrespected the mountain, causing the earthquake. If correct, the implication is that earthquakes (and perhaps other natural disasters) can be prevented if we could stop tourists from stripping.

This chapter gives a more prosaic explanation for earthquakes, as well as tsunamis, volcanoes, and hurricanes. Although it is possible to have earthquakes without tsunamis and vice versa, we will see that the biggest earthquakes and tsunamis are usually linked. Unfortunately, keeping out naked tourists is unlikely to have much effect on either one, meaning that we have to do a better job of making our infrastructure more resilient to such events.

In the media and public perception, natural and human-made disasters are often contrasted. Hurricane Katrina and the great Sumatran earthquake and tsunami of 2004 are considered examples of the former; the Chernobyl nuclear accident and New York's Triangle Shirtwaist Factory fire of 1911 (Box 2.1), with obvious human culpability, are examples of the latter. I take a somewhat different view.

When natural phenomena such as hurricanes, tornadoes, earthquakes, and volcanic eruptions are called "natural disasters" or "acts of God," the implication is that affected countries and individuals are innocent victims, powerless to avoid their dismal fate, or worse, deserving of their fate because of some perceived affront to the divine order. In 1750, two earthquakes struck the city of London. The Bishop of London claimed that this was the result of the population's behavior, because a ribald new novel (*The History of Tom Jones*), published a year earlier by author Henry Fielding, was being widely read. The novel had been banned in Paris. No earthquakes had struck

BOX 2.1 **The Triangle Shirtwaist Factory fire**

A shirtwaist is an old-fashioned name for a woman's blouse, and lots of them were manufactured in New York in the early part of the 20th century. In 1911, one of the deadliest industrial accidents in US history occurred at the factory of the Triangle Shirtwaist Company in Greenwich Village. The factory occupied the top three floors of a ten-story building. When a fire broke out in a scraps pile, workers were unable to escape because exit doors were locked, apparently to reduce the risk of theft. Managers with keys fled the fire first, dooming the workers left behind. There were no fire extinguishers, and fire department ladders were too short to reach the 9th floor of the building where most workers were trapped. A total of 146 people died, either from burns, smoke inhalation, or impacts with the street below (many workers jumped out of desperation to avoid the flames). The owners of the business eventually received an insurance payment equivalent to $400 per victim, of which they eventually paid $75 per victim as compensation to victims' families (the owners pocketed the rest). The accident spurred improvements in workplace safety regulation, helped establish several unions, and led to the formation of the American Society of Safety Engineers, which recently celebrated its 100th anniversary. A strikingly similar accident happened in 2012 in Bangladesh, which manufactures clothing for Western retailers. Since 2006, more than 1,000 Bangladeshi garment workers have died in factory fires, building failures, and related accidents.

BOX 2.1 **(cont.)**

Paris, proof positive of divine intervention and God's wrath with the lascivious English. The Malaysian government's finger-pointing after the 2015 earthquake in Sabah shows that such backwards views still persist.

Terms such as "natural disaster" or "act of God" are misleading, and can lead to some of the problems described in this book. These terms imply an element of malign unpredictability, where the black hand of fate or a stern, punishing Supreme Being strikes an innocent and unknowing populace. In this view, natural disasters are unavoidable, a dark part of the human condition, or worse, can be blamed on some innocent segment of the population, often a minority. The term "act of God" is actually enshrined in modern legal usage in many countries, and is sometimes used by insurance companies to avoid hefty insurance payouts or to justify higher premiums.

Such views may have been appropriate in the Middle Ages, when life spans were short, ignorance and superstition were widespread, books and literacy were rare, and natural science was a subject for religious persecution rather than a course of study in college. It is not clear why such views persist today. My feeling is that many disasters, whether the natural kind or human-made ones like the Triangle Shirtwaist fire, are the result of poor science or engineering, poor planning or management, or poor maintenance and operations. In this view, the Triangle Shirtwaist fire in New York and the Fukushima nuclear disaster in Japan have several features in common: Both were preventable, and both suffered from poor facility design, poor management, and lack of oversight. In both cases, common sense advice that in retrospect seems obvious was not followed. In the Shirtwaist example, make sure you have a fire escape, and don't lock employees in a building. In the case of Fukushima, don't put all your back-up power "eggs" in a low elevation "basket," especially in a low-lying coastal zone.

Table 2.1 lists the top 10 disasters in the US, both natural and human-caused, ranked by fatalities. Costs for the three most recent disasters are also estimated. Even in the US, a country with a history of collecting good statistical data, it's difficult to calculate costs rigorously, for several reasons. For example, when do you stop counting? Should future lost income be included? Should the cost of the war in Afghanistan, which the US waged for a decade following the 2001 terrorist attacks on the World Trade Center and the Pentagon, be included when calculating the overall cost of this disaster? Costs

Table 2.1 *Top ten disasters in US history, ranked by fatalities**

| Year | Type | Property Damage ($US) | Fatalities | Location |
|---|---|---|---|---|
| 1900 | Hurricane | – | 8,000 | Galveston, Texas |
| 1906 | Earthquake | – | 4,500 | San Francisco, California |
| 1928 | Hurricane | – | 4,000 | Florida, Puerto Rico |
| 2001 | Terrorist attack | $10B | 3,000 | New York, New York |
| 1941 | Military attack | – | 2,500 | Pearl Harbor, Hawaii |
| 1889 | Dam burst | – | 2,200 | Johnstown, Pennsylvania |
| 1893 | Hurricane | – | 2,000 | Louisiana |
| 2005 | Hurricane | $100B | 1,800 | New Orleans, Louisiana |
| 1865 | Ship wreck | – | 1,700 | Memphis, Tennessee |
| 1980 | Heat wave | $1B | 1,700 | Central, southern US |

* Notes for Table 2.1:

1. Disasters include both natural and human-made events.
2. Fatality estimates are rounded to the nearest 500 for some older, less well-documented events, otherwise to the nearest 100.
3. Cost estimates are accurate to about ±50%.

associated with earlier disasters are especially challenging to calculate. In contrast, the number of immediate fatalities can usually be estimated reasonably well.

Any such listing has a problem I call "table bias": Constructing the table inevitably involves over-simplification, because of the need to define arbitrary categories. For example, if we ranked disasters by cost as a percentage of GDP instead of fatalities, the great Mississippi River flood of 1927, which affected a third of the US, would have been near the top of the list. Should we consider a slowly evolving disease epidemic as a disaster? Rapid onset events tend to grab our attention, but the 1918–1920 influenza pandemic (Figure 2.1) killed more than half a million people in the US, and more than 50 million people worldwide. While it is far more serious than anything listed in

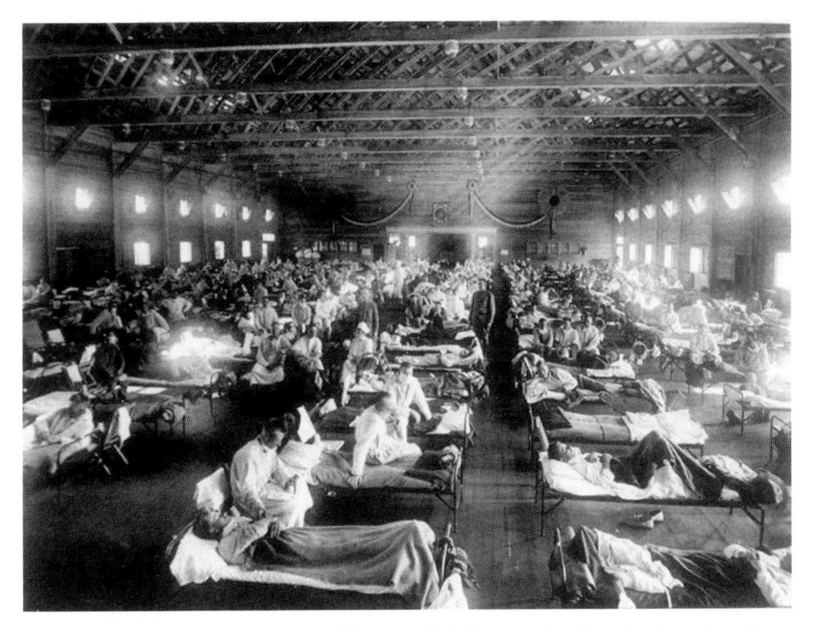

FIGURE 2.1 American soldiers with influenza in a hospital ward at Camp Funston, Kansas, close to where the virus was first reported in the US, in 1918. The global pandemic was probably caused by a virus similar to the modern H1N1 virus. US Army photograph, provided through Wikipedia Commons.

Table 2.1, it is usually excluded from natural disaster lists because it transpired over several years.

Different countries would have very different "top ten" lists. Italy experienced approximately 100,000 fatalities due to the 1908 Messina earthquake and tsunami. France suffered nearly 15,000 fatalities from a heat wave in the summer of 2003. However, listing by countries also has drawbacks, for example obscuring the impact of disasters that span borders. The death toll across Western Europe from the 2003 heat wave was at least 35,000 and may have exceeded 70,000, a clear indicator of the event's scale and impact. Temperature at London's Heathrow Airport was the highest ever recorded at that location. More than 200,000 fatalities were associated with the 2004 Sumatra earthquake, mainly due to the subsequent tsunami. Losses were spread across a dozen countries bordering the Indian Ocean, although Indonesia, Sri Lanka, India, and Thailand were especially hard hit.

In his book on global catastrophes, Eric Chaline includes slavery and several genocides of the 20th Century in his top-ten list. While these unfortunate episodes resulted in huge numbers of fatalities, I exclude them here because they represent deliberate malign choices made by society. Instead I focus on failures of infrastructure based on ignorance or neglect, with the implied optimistic assumption that once informed, we can do better.

Despite the various shortcomings, we can still draw useful generalizations from Table 2.1. One thing that stands out is that natural phenomena have been more dangerous to the US than military or terrorist attacks. This is not true for all countries – depending on time frame, a military attack or other human-made calamity can certainly dominate. If we focused on the former Soviet Union in the 20th century, that country's 25 million fatalities in World War II, or the 3 million fatalities associated with the Ukrainian famine of 1932–1933, would eclipse by far the effects of any natural disaster (the famine is generally ascribed to political and economic mismanagement by the Soviet Union rather than natural causes).

For most countries including the US, the human and financial costs of natural disasters are significant. Given these costs, it is surprising that countries do not invest more in disaster resilience – especially for earthquakes, tsunamis, and hurricanes, the most common ones. Going back to Table 2.1, if we assume that the recent US-led wars in Iraq and Afghanistan are a direct response to the terrorist attacks of 2001, and we assume that the long-term costs of these wars and other security-related changes to US society are of the order of $1 trillion (US) (a conservative estimate), then the US has invested roughly 100 times more in the responses to the attack than the actual cost of the attack (~$10 billion; Table 2.1). If this formula were applied to hurricanes, the US should be investing $15 trillion in infrastructure upgrades to flood-prone coastal areas in the Gulf of Mexico and Eastern seaboard, since the cost of storms in the last decade is at least $150 billion ($100 billion for Hurricane Katrina in 2005, plus $50 billion in 2012 for Hurricane Sandy).

A second generalization from Table 2.1 is that hurricanes are the most common entry. This turns out to be true for many countries, especially those located in tropical and sub-tropical regions. The fatalities and financial losses from hurricanes mainly reflect flooding from storm surge. These losses will likely increase in the future as sea-level rise and coastal development continue.

A third feature of Table 2.1 is that heat wave is a recent entry. This has several explanations, including a population that is increasingly elderly (the elderly are more vulnerable) and urban. Cities tend to be warmer than rural areas due to a phenomenon known as the "heat island" effect: Absorption of heat from the sun by dark asphalt and emission of heat by air conditioners and automobile engines raises the temperature of cities by several degrees compared to adjacent rural areas. However, global warming is almost certainly a contributing factor (Chapter 8). Heat waves are already starting to impact many countries and are likely to increase over the next few decades as a source of significant fatalities.

Note that volcanic eruptions are not listed in Table 2.1. Unlike earthquakes, volcanoes tend to give off warning signals prior to big eruptions. Although the US experienced a major eruption during the time period covered by the list (the 1989 eruption of Mt. St. Helens), there was plenty of warning thanks to monitoring by the US Geological Survey. Monitoring enabled timely evacuation, minimizing loss of life. Also, the eruption occurred in a relatively remote area, so property and economic losses were minimal. This could certainly be different in other countries. A significant fraction of El Salvador's land area consists of active volcanoes, and a major eruption there is likely to have a big impact. The Philippines experienced huge economic losses associated with the 1991 eruption of Mt. Pinatubo, although pre-eruption monitoring and timely evacuation minimized loss of life. A global table that spanned a longer period of time would probably include volcanic eruptions. Rare but

Table 2.2 *Top ten global disasters since 1970, ranked by fatalities*

| Year | Type | Fatalities | Location |
|---|---|---|---|
| 1970 | Storm, flood | 300,000 | Bangladesh |
| 1976 | Earthquake | 255,000 | China |
| 2010 | Earthquake | 222,500 | Haiti |
| 2004 | Earthquake, tsunami | 222,000 | Indian Ocean |
| 2008 | Cyclone Nargis | 138,300 | Myanmar (Burma) |
| 1991 | Cyclone Gorky | 138,000 | Bangladesh |
| 2008 | Earthquake | 87,500 | China |
| 2005 | Earthquake | 73,300 | Pakistan, India |
| 1970 | Earthquake | 66,000 | Peru |
| 2010 | Heat wave | 55,600 | Russia |

extremely large volcanic eruptions have even been implicated in global extinction events (see Box 2.5).

Accurate statistics on natural disasters for many countries are starting to become available and confirm that the general features of Table 2.1 are global in nature. For example, in 2013, the top three global disasters ranked by number of fatalities were Typhoon Haiyan (~6,000 fatalities, mainly in the Philippines), flooding in India (~5,500 fatalities), and a heat wave in the UK (760 fatalities), according to Munich Re. All three of these disasters likely have some relation to global warming. Swiss Re, another global re-insurance firm, has compiled a list of global natural disasters going back to 1970 (Table 2.2). As with the US compilation, tropical storm and cyclone-related storm surge is the biggest killer, followed closely by earthquakes, with heat wave a recent addition.

Dr. James Daniell, a researcher at the Karlsruhe Institute of Technology in Germany, has compiled a database on natural disasters from the start of the 20th Century (1900) until 2015. His data suggest similar generalizations: Flooding is the costliest natural disaster, accounting for about one-third of global losses, followed by earthquakes (about one-quarter).

In a sense, humanity is in the midst of a great transition, from being subject to nature's whims and having our buildings and cities periodically destroyed by natural phenomena we don't understand, to having a more resilient, robust urban infrastructure that can exist in harmony with the Earth. One hundred years ago, we knew very little about earthquakes and hurricanes, and certainly lacked technology to build structures that could withstand them. One hundred years in the future, it is likely that damage from earthquakes and hurricanes will be much less common. Perhaps this book can hasten that day.

## EARTH'S NATURAL CYCLES: VOLCANOES, CLIMATE, AND EARTHQUAKES

Volcanoes, climate, and earthquakes are usually explained separately in Earth science textbooks. I'm going to lump them together and attempt to convince you that they are actually connected in some important ways, at least on long time scales.

What do we know about the Earth and its inner workings? We know that the Earth has always been subject to cycles, changes, and violent events on a wide range of time scales. Most earthquakes and volcanoes are the inevitable consequence of motion of the Earth's plates, the dozen or so pieces of its hard outer shell that geologists call lithosphere. The Pacific Plate is the biggest. Most of its upper layer is called oceanic crust. Aside from a veneer of sediment on the top, oceanic crust is mostly basalt, a dark rock with lots of iron and magnesium. If you've ever visited Hawaii or Iceland, you've seen basalt. Both islands have a lot of it. Oceanic plates (and basalt) are created continuously at mid-ocean spreading centers, places on the sea floor where the Earth pulls apart and basalt wells up to fill the void. Like a conveyor belt, the oceanic crust moves away from the spreading center toward the edge of the ocean basin, getting older and accumulating an ever-thickening veneer of sediment that rains down from above. The sediment veneer includes shells from dead organisms, dust blown in from adjacent continents, and volcanic ash. In the oldest parts of the oceanic plate (far from its birthplace at the mid-ocean spreading center), this veneer can reach up to 1 km (more than half a mile) thick.

The Eurasian Plate (Europe, Asia, and some of the surrounding islands) is another big plate. Its upper layers are mainly continental crust – including rocks such as granite, the stuff of many upscale kitchen countertops – lighter in both color and density compared to basalt. This density difference will wind up being important when we talk about Earth's largest earthquakes and tsunamis (e.g. Box 2.4 and Figure 2.4). It also explains why the oceans are floored by oceanic crust – both continental and oceanic crust "float" on the layer of the Earth that lies beneath them (the mantle), but because of its higher density, oceanic crust sits at a lower elevation compared to continental crust (think about rubber duckies in a bathtub – the ones that leak sit lower in the bath because the water inside them has displaced some of the air, and water is heavier than air). The elevation difference between the surface of continents and the ocean floor (averaging about 4 km) defines the average depth of the oceans, since water flows downhill and fills Earth's low spots.

Figure 2.2 shows a map of the major plates and their motions. The length of the arrows indicates speed, and the orientation of the arrows indicates direction. Note the two arrows in Iceland pointing away from each other. This island nation sits astride the Mid-Atlantic Ridge, an example of a mid-ocean spreading center. Iceland is one of the few places on Earth where this feature is exposed above the ocean surface, and we can directly observe the process of spreading and formation of oceanic crust. Iceland gets wider at a rate of nearly 2 cm (three-quarters of an inch) per year because of this spreading process.

The motion of the plates is the ultimate cause of most volcanoes and earthquakes, and is a response to Earth's inner heat and resulting convection. Plate motion can be considered part of Earth's natural rhythm. It may even be a necessary condition for a planet to evolve and sustain life. Volcanic eruptions replenish Earth's surface water, help to maintain the balance of $CO_2$ in the atmosphere, and produce

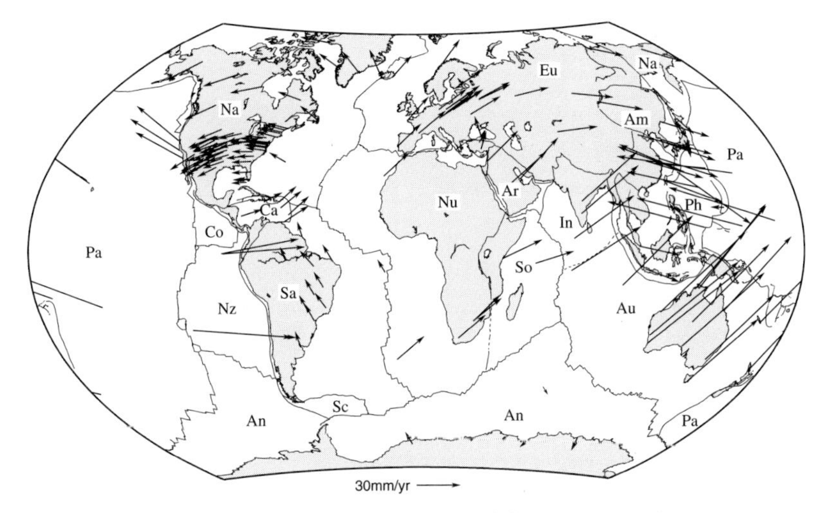


FIGURE 2.2 The Earth's major plates and their motions. The arrows show the direction of plate movement, and their length indicates speed. Plates that have significant continental crust include the Eurasian Plate (Eu), the North American Plate (Na) and the South American Plate (Sa). The Pacific (Pa), Cocos (Co) and Nazca (Nz) plates are mainly formed of oceanic crust and lithosphere. Modified from Sella et al. (2002).

a light dusting of volcanic ash on the Earth's surface. Without this, many soils would be less fertile. Ash from big eruptions can circle the globe several times before settling; this material is even found in ice cores in Greenland, far from active volcanoes, and is used to date the ice cores when studying Earth's past climate.

The process of subduction, where one geologic plate is pushed beneath another, creates deep trenches along many ocean boundaries, including the Pacific "Ring Of Fire" (Figure 2.3), and is the type of plate motion most often associated with catastrophic volcanic eruptions and with Earth's largest earthquakes and tsunamis. Most of the time, it involves an oceanic plate (e.g. the Pacific plate) that subducts beneath a continental plate (e.g. North or South America).

Reaching a trench is a death sentence for an oceanic plate, and they all die young, at least in geological terms. The oldest oceanic

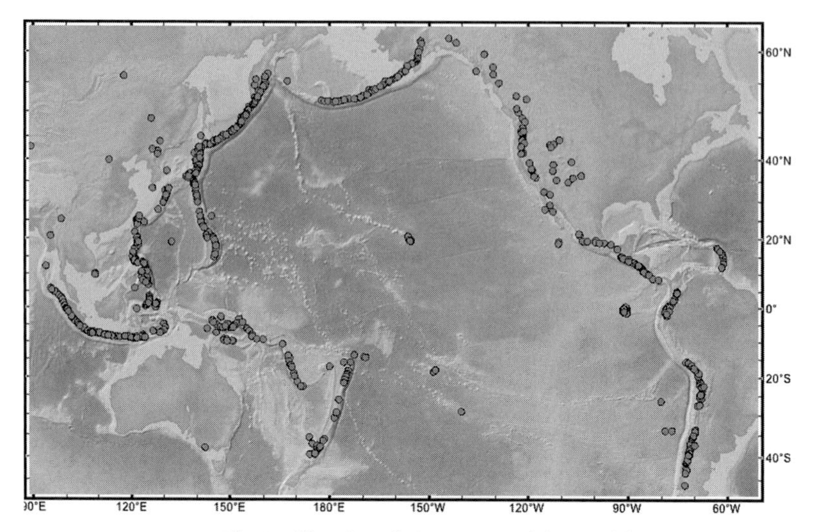


FIGURE 2.3 **The Pacific Ring of Fire.** Many of the world's active volcanoes (shown here with small circles) lay on or near the Pacific Rim, hence the name "ring of fire." A map of the world's major earthquakes would look similar. These volcanoes and earthquakes are a consequence of the geological process of subduction, which also causes Earth's deep ocean trenches (Figure 2.4). Figure made with GeoMapApp.

crust is only about 180 million years old, quite young compared to the age of the Earth (4.5 billion years old). From the time of its "birth" at a mid-ocean ridge to its subsequent journey and "aging" as it moves across the ocean floor before finally disappearing at the trench, the oceanic crust will soak up a lot of seawater (the water leaks into cracks in the rock, which reacts to form hydrated minerals). During subduction, the hydrous minerals are broken down, and the oceanic crust is forced to give back its excess water, like a sponge being squeezed. The extracted fluid contributes to (and in some ways causes) the process of volcanism. Without this constant squeezing and flux of water back to the surface (partly through volcanoes, Figure 2.4), Earth would lose its oceans in a few hundred million years. In a very real sense, volcanism on the ocean margins helps to maintain Earth's long-term water balance.

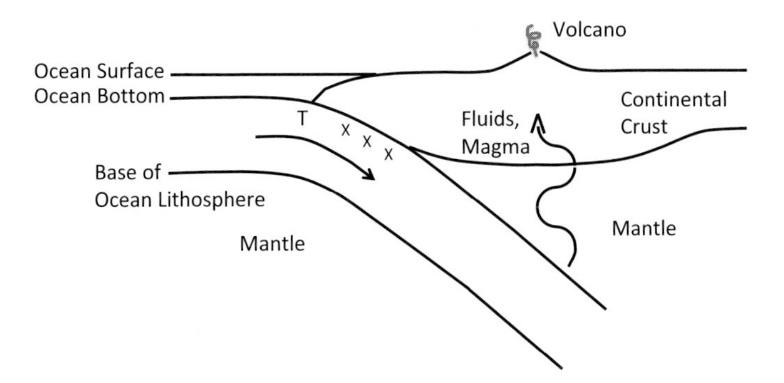


FIGURE 2.4 Cross section of a subduction zone, explaining the location of many earthquakes and volcanoes. "Continental crust" could represent the Canadian province of British Columbia, the US states of Washington and Oregon, or countries in Central America (e.g. Honduras) or South America (Colombia, Ecuador, Peru or Chile). "T" marks the location of the trench, a deep part of the ocean. The oceanic lithosphere subducts (is pushed and pulled) beneath the edge of the continent, in the direction of the arrow. Major earthquakes occur along part of the dipping boundary between oceanic lithosphere and the continent, marked by three X's (see also Box 2.4). As the subducting lithosphere moves down, it is squeezed by high pressure and releases fluids, causing melting of the mantle to form magma (molten rock) and volcanoes, returning large amounts of $CO_2$ and water to Earth's atmosphere.

The reason for the oceanic plate's death sentence as it reaches the trench is the density difference between oceanic and continental crust, about 6–7 percent. The oceanic plate loses the war with gravity as it reaches the edge of the ocean basin and runs into the bounding continental plate. Density is destiny in plate tectonics: The lighter stuff (continental crust) usually wins. The heavier oceanic plate is pushed beneath the edge of the lighter continental plate, which in the eastern Pacific is the western edge of North, Central, and South America; in the western Pacific it's the Eurasian Plate. The process does not go easily (I once heard it compared to trying to push a table under a door), so perhaps it's not surprising that big earthquakes and tsunamis happen here, marking the death throes of the oceanic plate.

People sometimes ask, "Where do plates go after they subduct (Figure 2.4)?" Obviously they go down, but how far? It's a good question, and the subject of ongoing research. Some are thought to go down all the way to the core–mantle boundary, roughly half way between the surface and the center of Earth, where they gradually warm up to ambient conditions and lose their integrity. But even here they may continue to influence events, hundreds of millions of years after they first began their downward journey. Their chemical make-up gives them a bit more of the "juice" (radioactive elements such as uranium and thorium) that leads to heating and volcanism, so it's possible they contribute to "hot spot" volcanoes (volcanoes like Hawaii that are far from plate boundaries), long after their disappearance from Earth's surface.

## HOW DOES THE MOTION OF PLATES LEAD TO EARTHQUAKES AND TSUNAMIS?

The answer has to do with faults (fractures in the Earth's crust that accommodate motion between the plates), a frictional process called "stick-slip," and a now widely accepted theory of earthquakes called "elastic rebound" (Box 2.2), first articulated more than a century ago by geologist H. F. Reid after the great San Francisco earthquake of 1906. His 1910 publication was a watershed event in Earth science.

Faults do not move steadily in response to the constant slow motion of the plates. Instead, frictional resistance along the fault causes the edges of the plates to get "stuck." Eventually the stresses build to the point where friction is overcome, and the rocks slip rapidly, resulting in the earthquake.

Figure 2.5 shows the offset of a fence built across the San Andreas Fault before it ruptured in the great 1906 earthquake. Observations like this helped Reid develop his elastic rebound theory. We now know that earthquakes are caused by a process of elastic strain accumulation and release, analogous to the slow stretching of an elastic band (the strain accumulation phase) until it reaches the point of breakage (the strain release phase). The strain accumulation phase may take hundreds of years, as strain builds up more or less at the rate of plate motion (perhaps a few centimeters per year).

### BOX 2.2 **Stick-slip and elastic rebound**

Plate motion leads to earthquakes on faults by a process called "stick-slip." Stick–slip is a surprisingly common process, the result of frictional resistance to motion. If you've dragged a chair across a hard floor and made a loud sound, you've experienced stick-slip. Try the following experiment on a concrete floor. Get a heavy brick or concrete block, and connect it to a bungee cord or other strong elastic cord. Gently stretch the cord, using a steady pull. At first, the cord stretches (strains), but nothing happens to the brick; it stays put, due to its weight and friction. However, if you continue stretching the cord, eventually the brick will jump forward.
The initial phase of this experiment (the "stick," when the brick sticks to the floor) represents the long (hundreds of years) pre-earthquake process of elastic strain accumulation on a fault, equivalent to the buildup of elastic strain energy in the bungee cord. The second phase of the experiment (the "slip") represents the earthquake: The brick jumps forward, rapidly releasing the energy stored in the elastic cord. What has happened is that the cord eventually exerted enough force on the brick to overcome the friction between the brick and the floor. This explanation for earthquakes is known as the elastic rebound model, and it has been around for more than 100 years.

In contrast, the strain release phase (the earthquake) can result in several meters of motion (slip on the fault) occurring in a few seconds to minutes.

Box 2.3 shows the cycle for a strike-slip fault, such as the San Andreas Fault in California. We imagine a fence built across the fault at right angles to illustrate the bending of the Earth's crust. The bending reflects the elastic strain, analogous to the stretched elastic band. Eventually, the elastic limit is reached, and rocks near the fault rupture, quickly releasing the stresses and strains that have built up over decades or centuries; that rapid release constitutes the

FIGURE 2.5 Fence in California offset by the 1906 San Francisco earthquake. Photograph by G. K. Gilbert.

earthquake. The resulting ground-shaking associated with abrupt movement of the rocks can be felt a long way from the fault. The earthquake displacements (the "slip") are usually visible at the surface, such as the fence offset in Figure 2.5.

## BOX 2.3 Stick-slip and elastic rebound on a vertical fault

Birds eye view of strain accumulation and release over three phases of the earthquake cycle, spanning many decades. This example is for a vertical strike-slip fault, the kind that ruptured in the 1906 San Francisco earthquake in California. **Left-hand side (1850):** A fence is built across the fault many years before the earthquake. **Middle (1905, one year before the earthquake):** The fence has slowly deformed (strained) over the previous 55 years, reflecting movement of the plate far from the fault, and bending the rocks near the fault because the fault is stuck (locked). **Right-hand side (1907, one year after the earthquake):** The fence as it appears shortly after the earthquake, prior to repairs. Rocks (and fence) near the fault have slipped or "caught up" with the far-field motion. Compare the right-hand panel to the photograph in Figure 2.5.

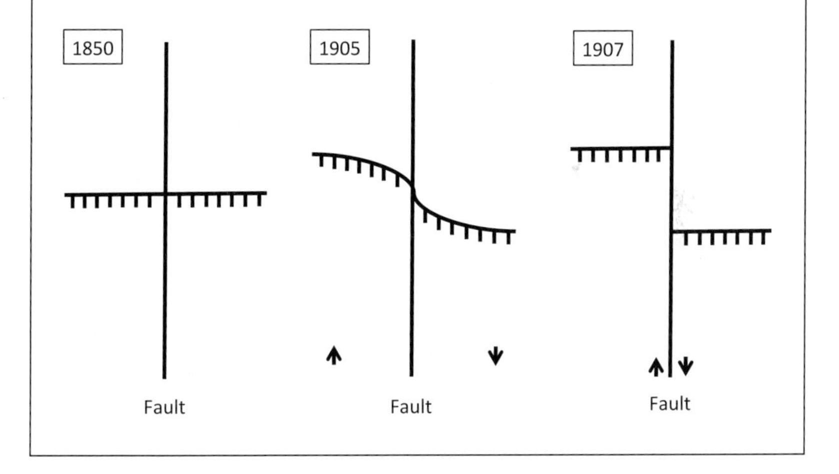


Box 2.4 shows the analogous process of elastic strain accumulation and release for a subduction zone fault. These faults tend to generate Earth's biggest earthquakes and tsunamis. During a subduction zone earthquake, the leading edge of the upper plate

flips up. If the rupture extends all the way to the trench, a large mass of ocean water is quickly accelerated upward. This bulge of water propagates across the ocean at high speed, making a tsunami wave that can cross the ocean in a few hours. Coastal areas close to the earthquake may only have a few minutes warning before the wave hits.

### BOX 2.4 **Stick-slip and elastic rebound on a subduction zone fault**

Three cross sections at a seismically active ocean-continent boundary (e.g. the west coast of Mexico or Central America) show the seismic cycle at a subduction zone. The process of elastic strain accumulation (stick) and release (slip) causes subduction zones to produce large earthquakes and tsunamis. The cycle can span several hundred years or longer. **A (top panel – pre-seismic or interseismic stage):** The down-going plate (oceanic lithosphere, left hand side) is pushed beneath the upper plate (commonly a continent); the boundary between them is a fault. Friction on the fault causes the leading edge of the upper plate to be dragged down by the subducting plate, contributing to the formation of a deep trench between them, marked by "T." B (**middle panel – seismic stage):** Eventually, the elastic strain builds to a point where frictional resistance on the fault is overcome, and the upper plate abruptly snaps back (ruptures). The leading edge (the underwater part) moves outward and upward, while inland portions subside. The abrupt motion stimulates damaging earthquake waves, and if the rupture reaches the surface, as it did in Sumatra in 2004 and Japan in 2011, it accelerates a large volume of ocean water (greatly exaggerated here), causing a tsunami wave that can propagate across the ocean. **C (bottom panel – postseismic stage):** Strain has been released, and the cycle begins again.

BOX 2.4 **(cont.)**

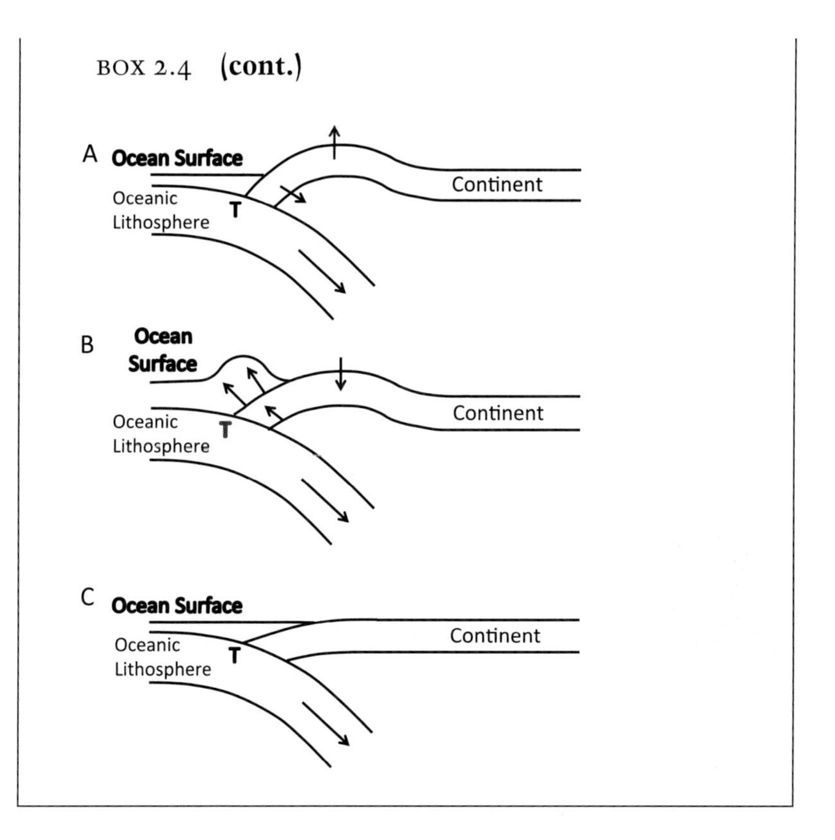


## VOLCANOES AND CLIMATE

Subduction-related volcanism is also part of a complex process that helps to moderate Earth's climate. Weathering of old mountain chains and subsequent movement of sediments downhill into the ocean removes carbon dioxide from the atmosphere by forming carbonate sediments (the stuff that makes corals and seashells). But we need some $CO_2$ in the atmosphere to warm the planet via greenhouse processes (see Chapter 8) otherwise the planet would be too cold. Volcanism usually returns just enough $CO_2$ to maintain atmospheric balance, sometimes called the "Goldilocks effect." The term applies to many natural processes that maintain a balance, and is from the fairy tale *Goldilocks and the Three Bears*, often attributed to British author Robert Southey. It's about a girl who intrudes in the house of three bears (mother, father, and baby) and rudely samples their various belonging, including trying to sit in their respective chairs (one too big, one too small, and one "just right") and sampling their porridge (too hot, too cold, and "just right"). Subduction zones are sometimes called Earth's factories, taking in raw material at the trench (basalt, sediment, and water) and later releasing modified material including $CO_2$, water, lava, and volcanic ash rich in plant nutrients, keeping our planet "just right" (Figure 2.4). The link between volcanoes, $CO_2$, and climate is explored more fully in the online appendix ("Exercises for Students").

There is a bit of circular reasoning in the above arguments: Life has probably evolved to deal with ambient conditions on our planet, whatever they are, so of course its going to seem just right to us, but there are some fundamental reasons to think the balance we have inherited is a pretty good one. Hopefully we won't be foolish enough to mess it up (Chapters 6 and 8 describe some attempts to do so).

Most of the time, Earth's regulatory system for water, carbon dioxide, and other atmospheric and oceanic constituents works pretty well, but occasionally it goes awry. The Cretaceous extinction, around 60 million years ago, and the Permian extinction 250 million years ago are probably the best natural examples (Box 2.5). Many geologists

### BOX 2.5 The Cretaceous and Permian extinctions

Part of Earth's natural rhythm includes major cataclysms that cause a number of species to go extinct, providing opportunities for new species to develop. The fossil record is full of such extinctions and new beginnings. The fossil content of sedimentary rocks is how early geologists first figured out how old various sedimentary rock layers were, at least in a relative sense (sedimentary rocks are formed when sediments are deposited in water; they include

BOX 2.5 **(cont.)**

sandstone, limestone, and shale). Sedimentary rocks and their fossils also helped Darwin formulate his theory of evolution.

The Cretaceous-Tertiary boundary, 60 million years before the present, is well known and marks the end of the dinosaurs. It was caused at least in part by a major meteorite impact: Sedimentary layers from that period contain small amounts of Iridium, a rare element that is enriched in many meteorites. Major volcanic eruptions at about the same time in India (the Deccan Traps) may have also contributed to the extinction. Rocks below (older than) the layers with Iridium contain dinosaur fossils, but rocks above it do not, so we know this event marks (and probably caused) the end of the dinosaurs. Rocks above this layer contain fossils of mammals (our ancestors), marking the beginning of the Tertiary period of geologic time.

The largest extinction on Earth is known as the Permian extinction. It happened about 250 million years ago, marking the end of the Permian period, and was responsible for extinguishing roughly 90 percent of life on Earth. While the cause of this is still being debated, huge volcanic eruptions in Siberia 250 million years ago are almost certainly one of the culprits. This eruption dumped so much $CO_2$ and other gases and small particles into the atmosphere that the Earth's climate and ecosystem were disrupted for hundreds of thousands of years, making conditions both on land and in the ocean untenable for many of Earth's species. Some geologists working on the Permian extinction tend not to get too worried when they hear the current debates on global warming; Earth has fared far worse in the past and will survive. Whether humanity survives is another question. We should therefore not be too cavalier about our own $CO_2$ emissions (Chapter 8). We don't want to get anywhere near conditions associated with end-Permian time, when so much life was lost. We don't know how close the planet came to having all life extinguished 250 million years ago.

believe we are in the midst of another mass extinction of species, this time caused by humanity's rapid expansion and impact on the environment. In honor of our own role in this particular extinction, some scientists have dubbed the current geologic time period the "Anthropocene."

In summary, the Earth has always had volcanic eruptions and earthquakes, and since they tend to occur in the same places (mostly on the boundaries between the plates), it really should not come as a surprise when these events occur. Most of these swans are white rather than black. The disaster occurs not because Earth did something unexpected, but because humans chose to build in locations prone to disasters and failed to take obvious precautions in the design and construction of infrastructure. Hence, many so-called natural disasters are actually a natural process turned into a disaster for human beings by virtue of economic or technological limitations, ignorance, or carelessness.

## HURRICANES, FLOODING, AND SEA-LEVEL RISE

Hurricane-induced flooding and wind damage is another example of a natural process turned into a disaster by human folly. Hurricanes have also been around for a long time and, just like earthquakes and volcanoes, occur in well-defined places, mainly affecting coastal zones (Figure 2.6). They are one of the ways nature redistributes heat on a planet with extreme temperature differences between the equator and poles. When one part of the ocean gets too hot, it is liable to generate, or magnify, an atmospheric storm system, transferring energy (via warm, moist air) from ocean to atmosphere and moving that air to cooler, drier areas, usually inland, northward, or both. Hurricane season is in the summer, when the ocean surface is warmest. As discussed in Chapter 7, as the oceans experience long-term warming, we will likely see increased hurricane intensity, since there is more energy to transfer.

If you live in coastal areas of the major US Gulf states (Florida, Louisiana, or Texas), or in Central America, or in the western

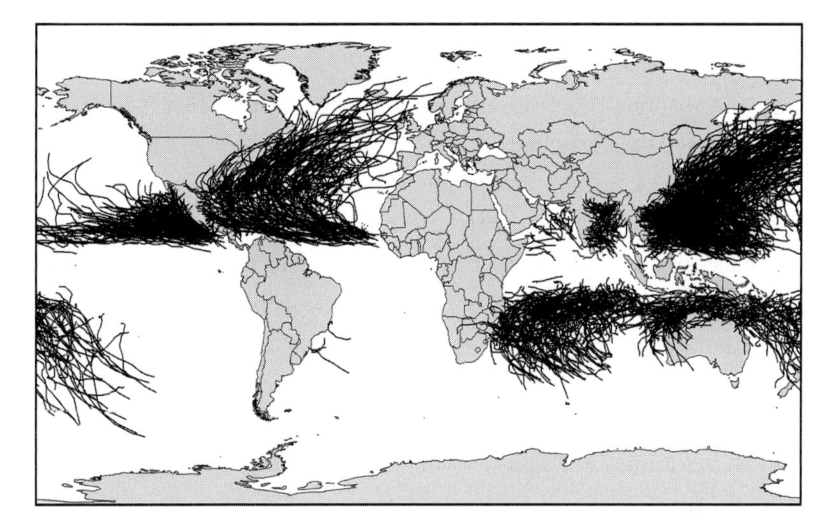

FIGURE 2.6 Named hurricane and tropical storm tracks since the year 2000. Note the large number of tracks north of the equator in the western Atlantic (Caribbean Basin, Gulf of Mexico, Eastern seaboard of the US and southeastern Canada) and western Pacific. Data courtesy of NOAA.

equatorial Pacific, it should not come as a surprise when a hurricane or typhoon strikes. They are frequent (often a dozen or more per year) and should be expected. The real surprise is the enormous damage to human structures these events continue to cause, the shock of public officials, the lack of strongly enforced building codes in regions frequented by hurricanes, and why television reporters continue to go out in the rain and wind to report on the damage.

There are two other puzzling aspects of hurricane-related disasters, at least in the US. First, business executives and politicians who normally consider themselves strict laissez-faire capitalists, disdainful of government interference, immediately call for declarations of disaster, government aid, and government-subsidized loans for rebuilding. Second, the construction industry lobbies furiously to dilute tough building codes when these are being considered by state legislatures, even though strong building codes should be good for the industry, generating more business and enhancing the reputation of a builder when a storm strikes and his or her building survives.

Until about 100 years ago, Florida, the US state that is probably most affected by hurricanes, had a set of natural defenses against the worst effects of hurricane flooding and storm surge: dense, fringing mangrove forests that reduced hurricane-related storm surge and limited construction in the immediate vicinity of the coast. That such vegetation also acted as critical nursery habitat for many species of fish, a mainstay of Florida's tourism industry, was an added benefit. Given all these benefits, you would think that mangroves would be common in Florida, but you would be wrong. In most parts of coastal Florida, including the highly developed southeastern coast from Palm

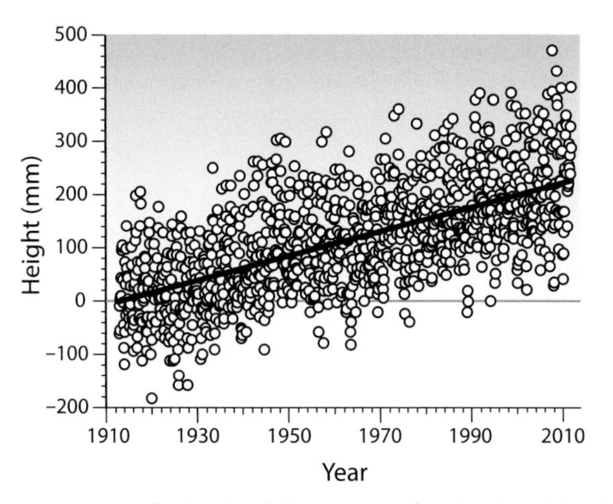


FIGURE 2.7 The height of the ocean surface (sea level) has been measured at Key West, Florida, since 1913. The measurement is made by a tide gauge, which functions much like a yard stick at the end of a dock, recording the instantaneous position of the ocean surface, but damping out wave action. This plots shows monthly averages, arbitrarily assuming that the height is zero in 1913. Since than, sea level has risen an average of about 200 millimeters (about 8 inches). A straight line fit to the data has a slope of 2.3 mm/yr (about 0.1 inch per year), indicating the rate of relative sea level rise. The ground on which the tide gauge is placed is believed to be stable, so the change represents an increase in absolute sea level that is probably related to global warming. Data courtesy of NOAA and PSMSL.

Beach to Miami, less than 1 percent of the original mangrove forest remains. Once the mangroves were chopped down, developers built sea walls, and backfilled the immediate area behind the wall, often using dredge material from the bay in front of the sea wall, further damaging the marine habitat. In this way the coastal elevation could be temporarily raised a few feet (~ 1 meter) above mean sea level, allowing construction of hotels, houses, and office buildings with great views. One difficulty with this approach is that, over time, the surface of this artificial fill subsides as the material compacts, lowering the land surface. Along with sea-level rise (Figure 2.7), this means that many developed coastal areas are now vulnerable to flooding, especially during tropical storms and hurricanes. The next chapter discusses some of the reasons for this state of affairs. Chapter 7 gives examples of flood disasters and shows how future costs from flooding might be reduced.

# 3 If We Know So Much about Natural Disasters, Why Do We Remain Vulnerable?

When it comes to the study of earthquakes, hurricanes, or other natural phenomena, there are several common factors that contribute to our susceptibility to disaster. In this chapter, I'll look at some specific concepts that relate to disaster vulnerability and preparedness in the context of our three themes – communication, time scale, and cost:

- how scientific uncertainty is communicated to the public;
- the problem of "sticky infrastructure" as it relates to the time scales of natural phenomena and the built environment; and
- building codes and the related issue of cost versus benefit.

## SCIENTIFIC UNCERTAINTY

Sometimes our scientific understanding is weak or incomplete. For example, an earthquake may occur in an unexpected region. Usually earthquakes occur on or near known active faults and plate boundaries – but not always. The 2001 Bhuj earthquake in western India is a good example. This magnitude 7.6 earthquake was responsible for nearly 20,000 casualties. Although earthquakes had occurred in the region in historical times, the area is far from an active plate boundary, which in this region is where the Indian Plate is moving north and colliding with the Eurasian Plate (Figure 3.1). That collision is responsible for the Himalayas. The situation is similar to the example sketched in Figure 2.4, except instead of an oceanic plate like the Pacific subducting beneath the leading edge of a continental plate, two continental plates collide (Figure 3.2). Since both plates are continental, and therefore have similar densities, neither wants to subduct: The resulting collision deforms and

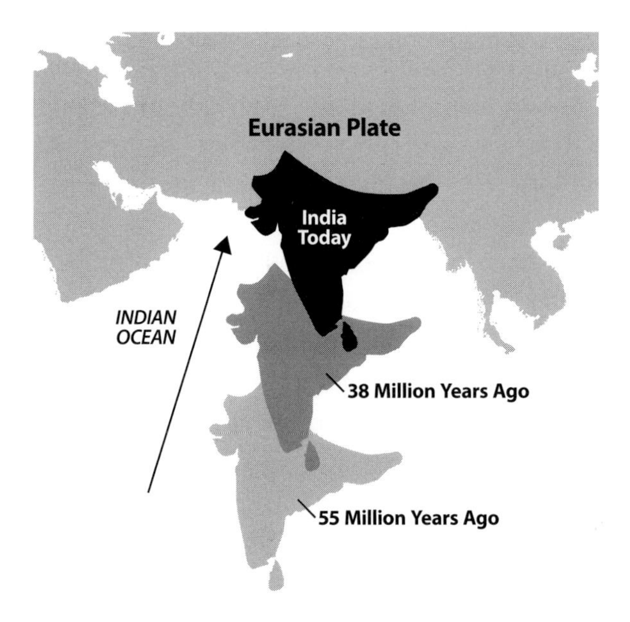


FIGURE 3.1 India was not always part of Asia. For the last 70 million years, the Indian Plate has been migrating northward, and about 40 or 50 million years ago began to collide with the Eurasian Plate. The resulting collision formed the Himalayas along India's northern boundary (see Figure 3.2).

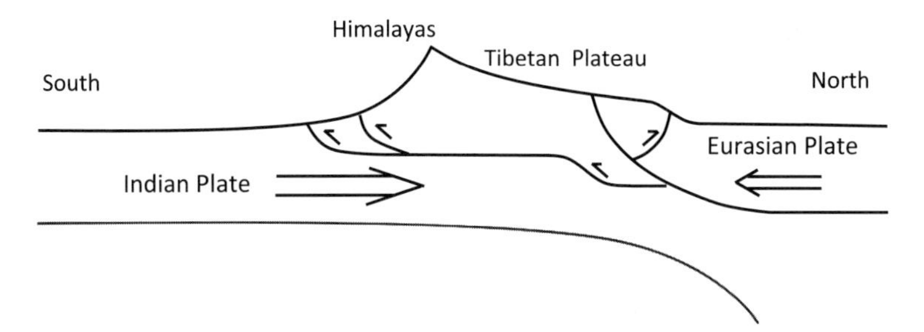


FIGURE 3.2 Example of a continent-continent collision zone. Because continental crust (India, Eurasia) is relatively light (less dense), it does not subduct easily. The leading edges of both plates crumple, with some of the crust pushed up, forming the Himalayas, while denser parts are pushed down. Curved lines with single-sided arrows are faults, showing how crustal blocks are pushed up to make the mountains. Instead of a single fault separating the plates, as in Figure 2.4, the plates here are separated by multiple faults defining a broad, complex boundary zone. Because there are so many faults (simplified to just a few in this diagram) it is more difficult to assess earthquake hazard.

pushes up continental crust at the leading edge of both plates, in this case forming the Himalayas. Geologists think a similar process formed the Appalachian Mountains in the Eastern US and the Alps in Europe. The resulting plate boundaries are complex, better described as broad plate boundary zones with dozens of active faults rather than a single discrete bounding fault.

The 2001 Bhuj earthquake is still being studied by scientists. The area had not been highlighted as one of exceptional risk, though perhaps in hindsight it should have been. A similar earthquake occurred in 1819, close to the 2001 event, indicating the importance of historical data and considering long time spans for assessing risk. The point is that, while science can do a pretty good job of identifying most risk areas, our knowledge is not perfect. There are going to be some black swans. But that should not be an excuse for inaction everywhere.

Weak or incomplete scientific understanding has two negative impacts. First, if scientists don't know the answer, they obviously can't give useful advice. More importantly, a wrong answer, or a failure to give a sense of the uncertainty in a scientific statement, undermines the credibility of the larger enterprise. If politicians or business people ask for advice, and then are given vague or wrong answers, they will be reluctant to repeat the process. This is rare, but it does happen.

The town of L'Aquila in central Italy is a tragic example. In 2009, a magnitude 6.3 earthquake killed more than 300 people. Prior to the earthquake, a series of small tremors had shaken the area, raising public concern. A commission of scientists and engineers was convened to assess the situation and provide guidance to the public and government officials. The commission met and shortly after stated that there was no unusual earthquake risk. Unfortunately, a major earthquake occurred soon after the commission had issued its statement, causing a furious backlash that led to legal action against the commission members, including charges of manslaughter. Seven members of the group were found guilty in 2012 by a lower court and sentenced to several years in jail. The convictions were reversed on appeal in 2014 for six of the commission members, while the seventh had his sentence reduced.

The case is troubling for several reasons. All experts agree that it is impossible to predict earthquakes with our current scientific understanding, and the legal proceedings could have a chilling effect on future public statements by scientists. On the other hand, some of the commission's statements do appear cavalier in hindsight. One government official, the deputy chief of Italy's Civil Protection Department, apparently advised worried residents to go home and sip a glass of wine, even specifying the type: "Absolutely a Montepulciano" (Pritchard, 2012). For students of wine, Montepulciano d'Abruzzo is the local red wine of the Abruzzo region of east-central Italy, where L'Aquila is located.

I am sympathetic to the commission members; some of their quotes were taken out of context, which is often a problem when discussing complex issues with the media. For example, the above quote was a direct response to a reporter who had specifically asked if residents should go home and have a glass of wine, and I could imagine myself responding in much the same way at the end of a long day of meetings.

Dr. David Spiegelhalter, Professor of Public Understanding of Risk at Cambridge University, said in the same BBC article, "Communication is at the bottom of this whole case" (qtd. in Pritchard, 2012). David Ropeik, writing in *Scientific American*, described the legal judgment as "not against science, but against a failure of science communication" (2012).

Let's look at this event in a little more detail. Part of the problem is that we (scientists) lack a good way of describing to the general public the hazards inherent in low probability but high impact events. Earthquakes with long recurrence times are a good example. It is also difficult to get public officials to pay attention to the possibility of such events before they happen, when there is still time to reinforce buildings and other infrastructure. As we will see in the next chapter, this was a factor at Fukushima.

The occurrence of a large number of small earthquakes in the months preceding the 2009 L'Aquila earthquake featured prominently in the trial and in media comments, but it may not actually be relevant. The commission was correct in downplaying the importance of this "seismic swarm," because this type of activity does not provide good clues to future behavior, at least with our current understanding. Sometimes seismic swarms lead to a big earthquake, perhaps because they "load" (add stress to) adjacent sections of a fault. But just as often, nothing happens, perhaps because the numerous small earthquakes have actually relieved some of the stress on the fault. Currently, our available instrumentation and scientific understanding make it difficult to distinguish between these possibilities. While there are a number of people claiming to be able to predict earthquakes, their predictions are usually announced after an event and lack scientific credibility.

The historical pattern of earthquakes in the region is interesting and relevant. Not counting the 2009 event, since 1349 AD there have been at least three devastating earthquakes in or near the L'Aquila region, roughly one every 220 years, with the last one in 1703. The 1703 event devastated the city because it was built on an ancient lakebed, which amplifies seismic shaking and ground acceleration. This can be hard on older, unreinforced masonry buildings. The 1703 event caused several thousand fatalities. Since it had been more than 300 years since the last large earthquake, perhaps the area was "overdue" for the 2009 event. A seismic hazard map published in 2003, based in part on such historical information, clearly showed the Abruzzo region to be located in a hazardous zone (Boschi, 2013).

Spiegelhalter is right that communication is a key part of the case. In particular, better communication of the uncertainty of any earthquake forecast would have been helpful. With the benefit of hindsight, perhaps a better public statement would have been to simply say that there is always a danger of major earthquakes in this area, and caution is therefore warranted.

One issue that I have not seen addressed in this case relates to the construction standards in L'Aquila. A relatively modest-sized earthquake like the 2009 event should not cause much damage or loss of life with modern construction techniques. L'Aquila has a number of older structures, including some medieval ones, but it is usually possible to re-engineer older buildings to withstand the modest level of shaking that occurs in magnitude 6.0–6.5 earthquakes at relatively low cost (larger earthquakes are a different matter, especially for medieval buildings where aesthetics become an issue). One of the commission members, Dr. Franco Barberi, had actually compiled a report on the safety of buildings in the region in 1999. Many buildings in L'Aquila were identified as being at moderate to high risk of failure in an earthquake. Government officials and civic leaders therefore had a decade after Barberi's report to strengthen the town's infrastructure, or at least publicize which buildings would be unsafe in the event of an earthquake so that residents could make informed decisions about where to live. While scientists could have done a better job of communicating the risk to the public, politicians could have done a better job of responding to that earlier study. The issue of the legal liability of officials who failed to act on Barberi's prescient report should also be considered in this case.

Stephen Hall has written an excellent article on this episode for readers interested in more background (Hall, 2011).

## TIME LAG, STICKY INFRASTRUCTURE, AND THE CHALLENGE OF RELOCATION

Another factor that explains our continuing susceptibility to disasters is related to the relatively late stage of our understanding of critical natural processes, combined with the "stickiness" of infrastructure: Once constructed, buildings and cities are difficult to move. Most cities were established long before we knew much about hurricanes or earthquakes, and it would not be economically feasible to relocate them now. For certain hazards, there was a time lag between establishment of infrastructure and our scientific understanding of that hazard.

Consider storm-related flooding in Florida: Our understanding of storm surge was incomplete when south Florida's coastal development was occurring. During Florida's first few building booms (there have been several since Julia Tuttle, often considered the founder of Miami, first convinced Henry Flagler to extend his railroad south to Miami and Key West), the ecological importance of mangrove forests was not appreciated the way it is today. The devastating power of hurricanes was not clear in Florida until the great storms of 1926 (Miami), 1928 (Lake Okechobee), and 1935 (Florida Keys) killed hundreds to thousands of people. The 1926 storm caused huge financial losses, affecting the region for several decades. The ecological and flood prevention benefits of mangroves are now well recognized, and they are currently protected by a number of laws.

San Francisco and Los Angeles comprise a significant fraction of California's population. Both are located close to the San Andreas Fault, part of the Pacific–North America plate boundary and an area now known to experience major earthquakes every few hundred years. San Francisco (established in the early 1800s) and Los Angeles (late 1800s) were already thriving metropolises when the 1906 San Francisco earthquake struck. Los Angeles will likely experience a devastating earthquake within the lifetime of our children or grandchildren. One section of the San Andreas Fault north and northwest of the city last experienced a major earthquake in 1857, and the recurrence interval for similar events is now thought to be about 200 years (Jackson et al., 1995). If earthquakes were exactly repeating (they are not) we might expect another big one on this section of the fault around 2060, but the uncertainty on this type of calculation is at least 50 years. Different sections of the fault also appear to have different recurrence intervals, and even the model of regular earthquake recurrence is questionable (e.g. Weldon et al., 2004, 2005). So while Los Angeles is already within the expectation window for a major earthquake, it is also true that nothing might happen until the year 2100 or even later.

San Francisco's 1906 earthquake was a major event in Earth science. Before 1906, geologists knew very little about earthquakes. We learned in Chapter 2 about geologist H. F. Reid who studied the event and came up with the elastic rebound theory of earthquakes. Reid measured the displacements of the Earth's crust associated with the earthquake and demonstrated the importance of careful geodetic measurements. (Geodesy measures the position of markers on the ground, or the lengths or angles between them.) However, it wasn't until the 1960s and 1970s that geologists came up with the concept of plate tectonics, related crustal blocks to plates, figured out how and why plates move, and identified plate boundaries as faults where earthquakes are likely. Modern satellite techniques can measure the slow motion of plates (e.g. Figure 2.2), and we now understand that the cumulative effect of this motion is directly related to the large fault offsets that happen during earthquakes, such as those measured by Reid more than a century ago.

Unfortunately this knowledge came long after San Francisco and Los Angeles were thriving metropolises. We now understand that these cities were built close to the San Andreas Fault, a major plate boundary likely to experience a devastating earthquake every few hundred years. Relocation is not an option for these cities or, indeed, most cities. Much the same applies to coastal cities affected by hurricanes: Our understanding of the great storms came long after our thriving coastal cities were established. Since we can't move our cities, we have to re-engineer our infrastructure. We need to build smarter, as described in the next section (The Importance of Building Codes).

Some farmers in Central America have evolved an interesting strategy for dealing with the hazards and benefits of volcanic eruptions. Farming has gone on for hundreds and possibly thousands of years here, and volcanoes are an intrinsic part of the landscape. Individual volcanoes may lie dormant for hundreds of years, but there are enough volcanoes (a few volcanoes erupt every few decades), that the possibility of eruption is appreciated by nearly everyone – there is community memory.

The volcanoes in Central America often erupt explosively, so most farmers don't live on the volcano. The lavas have lots of water, intrinsic to the subduction process (Figure 2.4), which can flash to steam as the magma moves higher in the Earth's crust, where pressure is lower. Much as a popcorn kernel pops when heated quickly (the water in the kernel expands to form steam), the magma can explode, forming volcanic ash. The ash breaks down quickly in the tropical climate, releasing nutrients. Volcanic ash often has relatively high concentrations of potash and phosphate, two critical plant nutrients.

Because of these nutrients, volcanoes turn out to be fertile places to grow crops, especially a few years after an eruption, when tropical rains have started to break down the volcanic ash and newly available nutrients are accessible to plants. Farmers can grow plants on this new ash for many years after an eruption without having to pay for fertilizer.

Communities along the banks of the Mississippi and Missouri rivers might be wise to follow the example of Central American farmers. Rather than having expensive infrastructure on the floodplain, which periodically experiences devastating spring floods, perhaps these areas could revert to green space or be used only for agriculture. The occasional high spring flood can be devastating for certain types of crops (farmers cannot get on the land early enough for spring planting), but in the long run, flooding is beneficial because silt deposits from the flood help to replenish the soil. In the US, a federal crop insurance program, which already exists to cover some weather-related disasters, could be modified to include this type of coverage, helping to smooth out good and bad years. US federal policy now promotes the opposite behavior. Taxpayer-funded levee construction encourages people to build expensive infrastructure in flood plains. The levees work for a while, giving people a false sense of security, then a high rainfall year occurs, leading to devastating floods that overrun levees and cause billions of dollars of damage. Even after such disasters, there is resistance to "rebuilding smarter." While some communities take the opportunity to move sensitive infrastructure to higher ground, others doggedly rebuild in the same flood-prone areas, encouraged by insurance schemes (often federally subsidized) that encourage foolish rebuilding strategies. Coastal flooding is discussed in detail in Chapter 7. Some possible solutions to the relocation problem are discussed in Chapter 9.

Government officials and others who study floods can inadvertently encourage bad behavior by using such phrases as "the 100 year flood" to describe an unusually high-flood year. To members of the public, this phrase implies that there is some predictability to flooding, or at least some statistical knowledge of the frequency of high-flood years. If you've just been hit with the hundred-year flood, perhaps it is safe to rebuild on the floodplain because you now have another 99 years before the next one, and the economic lifetime of many buildings is less than 50 years. Such talk is nonsense. Scientists and government officials actually have no idea of the future frequency of flooding. Even if future conditions remained constant, we would have to have several thousand years of accurate weather and flood data along an entire river system in order to get reasonably accurate statistics on the frequency of past flood events. Such data do not exist. As with earthquakes, a flood record based only on information from the last one or two hundred years is too short to provide reliable statistics. Even if we had such data, the resulting model would still have limited predictive value. The reason is that the record would reflect past climate conditions but not future conditions. Since the flood frequency and climate are changing (see Chapters 7 and 8), we can expect the frequency of future flooding to change as well. The term "100-year flood" is misleading, and should be avoided.

History and archeology hint at several examples of cities that were destroyed by volcanic eruption, earthquake, or flood, and were not rebuilt in the same location. Presumably the remaining inhabitants chose to build elsewhere. It is unlikely that this reflected a detailed analysis of the return period of the disaster. Perhaps for religious or health reasons survivors did not want to build on top of deceased former inhabitants. The Italian city of Pompeii is a well known example, destroyed by an eruption of Mt. Vesuvius in 79 AD.

While relocation in response to natural disaster has happened in the past, it will be more difficult in the future. In the first chapter of this book, I discussed the increasing costs of natural disasters and suggested that this was due at least in part to population increases (Figures 1.1 and 1.2). Population growth also bears on the issue of relocation. In an increasingly crowded world, there may not be enough new locations that are available or affordable for displaced populations who otherwise might like to relocate. This will be a problem for the populations of low-lying island nations (e.g. the Maldives in the Indian Ocean, with an average elevation less than 2 meters, or the Marshall Islands in the western Pacific) and poor coastal nations (e.g. Bangladesh). All three countries are already feeling the effects of sea-level rise. Typhoon-related flooding in Bangladesh after a few more decades of sea-level rise has the potential to produce millions of refugees. Our society does not have systems in place to deal with problems of this scale.

## THE IMPORTANCE OF BUILDING CODES

If we can't move cities away from fault or flood zones and other high-risk areas, what is the answer to reducing risk from natural phenomena? The primary solution is zoning codes (where to build) and building codes (how to build). I'll focus on the latter. Florida, especially Miami-Dade County, has rigorous building codes designed to reduce damage from tropical storms and hurricanes. These began to be developed soon after the deadly 1926 hurricane, one of the costliest on record (by some estimates it was comparable to the economic damage caused by Katrina). Hurricane Andrew in 1992 also caused widespread damage and resulted in further toughening of standards. California, Japan, New Zealand, and Chile lead the world in the development and rigorous enforcement of building codes designed to minimize losses of life and property from earthquakes. In California's case, much of the impetus came from the Long Beach earthquake of 1933, a moderate size (M 6.4) event that nevertheless caused widespread damage throughout Southern California, including a number of schools. The fault responsible for this event was the Newport-Inglewood Fault, which lies partly offshore. The fault is about 80 km southwest of and parallel to the San Andreas Fault, the major boundary fault separating the Pacific and North American plates.

The Long Beach earthquake struck at about 6 p.m. local time. Recognizing that the death toll would have been much higher had school been in session, the California Legislature passed the Field Act (named for Assembly member Charles Field, who pioneered the legislation) within a few months. The Field Act mandated strict building codes for schools and hospitals. In particular, this act dealt with the issue of unreinforced masonry buildings, which suffered abnormally high losses in the earthquake compared to other types of buildings.

One interesting aside is that the Long Beach earthquake was the first earthquake where accelerations of the ground during an earthquake were measured directly by using a new class of instrument, the accelerometer. Buildings can fail when they experience accelerations in excess of their design standard. Earthquake-induced ground accelerations that approach or exceed Earth's background gravitational acceleration (the thing that keeps us anchored to the ground) are considered large and dangerous, but are fortunately rare. Ground accelerations can be estimated from standard seismograph recordings, but these sensitive instruments may go off scale near the earthquake, where accelerations can be quite strong. This phenomenon is known as "clipping," or saturation (Box 3.1) and can affect many types of measurements. The accelerometer, also known as a strong motion sensor, is used both for building design studies (Box 3.2) and to give early warning of strong earthquake shaking.

### BOX 3.1 Drift, aliasing, and clipping

Scientists spend a lot of time figuring out how to measure things. A surprising number of problems in the natural hazards arena relate to measurements that are inadequate in some way. Whatever technique is used has to be both precise (meaning that the measurements are reproducible) and accurate (meaning that the measurements bear some relation to the truth). Drift refers to a problem when measurements become progressively more inaccurate with time; the instrument "drifts" out of calibration.

The measurement technique also has to adequately describe the signal of interest. Aliasing and clipping refer to problems when this last criterion is not met. Aliasing is a term from electrical engineering, used to describe a situation where an electronic device, such as a digital receiver or recorder, fails to sample the signal of interest frequently enough to capture reality; the resulting under-sampled signal is said to be "aliased." If you've ever seen wagon wheels that appear to go backwards in an old western movie, you've seen an example of aliasing (the frame rate on the camera is too slow to resolve the rotation). Earth scientists often generalize the term to indicate any set of observations that are inadequately sampled in time or space to describe the phenomenon of interest. For example, the time between major earthquakes or tsunamis may be hundreds of years. Knowing that time is clearly important for hazard assessment. For regularly recurring events, this is called the recurrence period. The period can be correctly inferred only if we catch all the major earthquakes within a representative span of time. In Chapter 2, I discussed the problem of estimating flood hazard based on records that only go back one or two hundred years. In Chapter 4, I'll discuss how failure to consider the need for a sufficiently long earthquake and tsunami record contributed to the Fukushima disaster; in effect, authorities were relying on an aliased record to assess risk. In Chapter 5, I'll look at how clever sampling has allowed geologists to get around the aliasing problem and piece together the long history of major earthquakes in the US Pacific Northwest, clarifying the earthquake and tsunami

BOX 3.1 **(cont.)**

hazard there. In Chapter 8, I'll consider some of the challenges in putting together a sufficiently long record of Earth surface temperature to evaluate global warming. The Appendix includes a sample problem on aliasing as it relates to earthquake hazard.

Clipping refers to missing the peaks or valleys of a signal, usually because the signal is too strong for the design of the instrument (think of cheap speakers that distort when the music gets too loud). The principle is shown in the figure above. The signal in this example is a simple wave with highs and lows, but the maximum signal is missed ("clipped") because the instrument is only capable of recording signals to the limit of the upper horizontal line. Similarly, the minimum signal size is missed because the instrument lacks sufficient sensitivity. Seismometers for measuring earth motion during an earthquake tend to clip at the upper end if they are located near the earthquake's source, where shaking is strong. This initially made it difficult to design buildings to survive earthquakes. Until strong motion sensors came along (basically, seismometers that don't clip large signals), engineers didn't know how strong the ground-shaking close to an earthquake actually was. Some of the radiation sensors deployed near the Fukushima nuclear plant in Japan similarly went off scale during the disaster, meaning that the maximum levels of radiation released during the accident are poorly known (see Chapter 4).

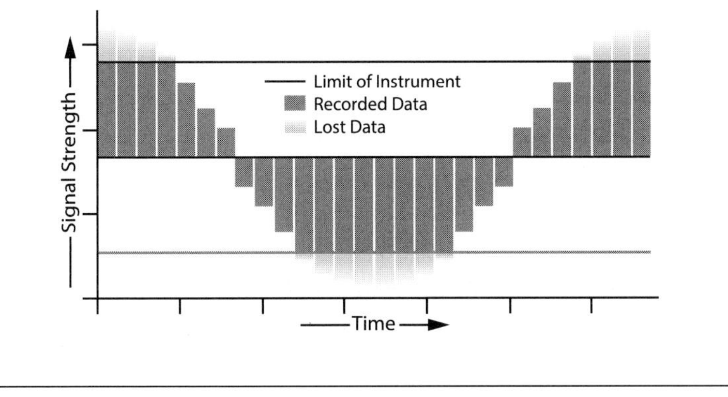

## BOX 3.2 The importance of acceleration

Vertical acceleration of the ground during earthquake shaking is an important factor to consider in building design and survival in an earthquake. To understand why, it's important to first understand what defines the strength of a building. This is a complex topic and includes things like choice of material, overall building design, and the manner in which the various building elements are constructed (shoddy construction can easily trump a good design). However, one factor is simply the weight of the building, the product of the building's mass times the acceleration due to gravity. This determines the frictional strength of some of the building's main structural elements, at least in the absence of earthquake-induced acceleration. Upward motion of the ground during an earthquake temporarily counteracts the effect of Earth's gravity, reducing the effective weight of the building and the effective frictional strength of some of the building elements. This is why unreinforced masonry is a bad idea; a significant part of its strength derives only from weight and friction – fine if there are no earthquakes but potentially deadly in the presence of earthquake shaking and vertical acceleration. Earthquake-resistant designs include lateral bracing and steel-reinforced masonry (e.g. rebar). These design elements do not lose their strength in the presence of upward acceleration. Horizontal accelerations are also important and are addressed through lateral bracing and other aspects of building design.

A simple example comes from Lincoln Logs, a popular children's toy (at least before the era of video games) where notched, miniature wooden logs are stacked to form model log cabins and other buildings. The models are reasonably stable as long as you don't shake the base. While moderately strong in the presence of sideways shaking (horizontal acceleration), even a mild up-down motion causes the structure to collapse. Imagine now that you could wire all the logs in the model together, analogous to the lateral bracing and steel reinforcement that is a key feature of earthquake-resistant construction. The model can now withstand even determined

BOX 3.2 **(cont.)**

attacks by siblings. Ironically, the toy was invented by John Lloyd Wright, son of famed architect Frank Lloyd Wright. Wright modeled the notched logs using an example from the Imperial Hotel in Tokyo, which was designed by Frank Lloyd Wright and said to be earthquake-proof by virtue of interlocking beams in its foundation.

The principle behind an early warning system for earthquakes is based on the fact that earthquake waves travel through the Earth much more slowly than radio signals through the atmosphere, giving several tens of seconds notice for some cities that are not directly on top of the earthquake. This may be sufficient time to shut down critical infrastructure such as fast-moving trains and gas pipelines, and for people to "shelter in place" (e.g. under a strong desk). Ironically the accelerations experienced at Long Beach exceeded even the design limit of the new "strong motion" instruments (they clipped), so only a minimum estimate was obtained. Estimates of maximum acceleration likely to be experienced at a given location are now used routinely by engineers when designing buildings for earthquake resistance.

Of course, building codes must be enforced if they are to be useful. California, Japan, New Zealand, and Chile all do a good job in this regard. While similar building codes have now been adopted in most countries exposed to seismic hazard, enforcement can be spotty. Corruption may play a role; in general there is a correlation between a country's level of corruption and its adherence to strict building codes. Corruption also correlates with poverty, and some people draw a link between the level of corruption in a country and that country's wealth, because of the deleterious effects on business (Figure 3.3). However, as generations of professors have told their students, correlation does not indicate causality. If it did, a wealthy country like Italy would not have a Corruption Perception Index (3.9) that is similar to some third world countries. Many factors contribute

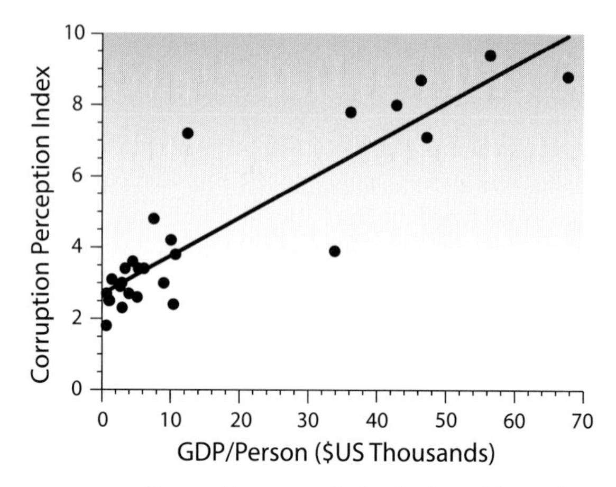


FIGURE 3.3 Corruption versus individual wealth for the nations mentioned in this book. Corruption is measured by the Corruption Perception Index (CPI), based on interviews with business people doing business in that country. CPI varies from 0 (very corrupt) to 10 (very clean). Wealth is measured by Gross Domestic Product per person. Data from Transparency International (www.transparency.org) and the United Nations; illustration prepared by the author.

to a country's wealth beyond the level of corruption, including the abundance of natural resources (lots of them make you wealthy) and history (lots of invasions by your neighbors make you poor), so we should not be too quick to make moral judgments on countries that don't adhere to strict building codes. Rather than corruption, it is just as likely that a poor country builds poorly because it lacks resources to do otherwise.

Many issues related to earthquake safety and poverty came to the fore in the aftermath of the disastrous January 2010 earthquake near Port au Prince, Haiti. This M 7.0 event was initially thought to have caused more than 200,000 fatalities, although more recent estimates are lower. A report published in December 2010 by Athena Kolbe and colleagues in the journal *Medicine, Conflict, Survival* estimated fatalities at approximately 159,000. A later unpublished report commissioned by USAID estimated fatalities between 46,000 and 85,000. Given the scale of damage, the true number may

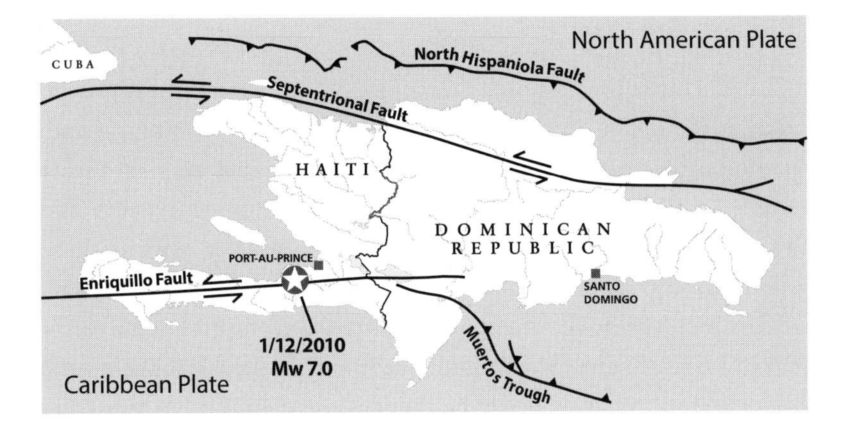


FIGURE 3.4 Plate tectonic setting of Haiti and the island of Hispaniola. The island lies within a complex boundary zone between the North American Plate (roughly the top half of the map) and the Caribbean Plate (roughly the bottom half of the map). A network of faults (only the major ones are shown here) separates the two plates. While earthquakes could occur on any of these faults, including ones not shown, large earthquakes are more likely on the major named faults: the Septentrional Fault to the north, and the Enriquillo Fault to the south, shown as solid lines, with arrows representing the sense of motion. These are vertically oriented faults, accommodating mainly sideways motion, and are unlikely to generate a large tsunami. Other faults, mainly offshore, are oriented at some angle from vertical and are shown with small triangles to indicate the dip direction. These faults accommodate some vertical motion and could generate a tsunami. The Enriquillo fault runs close to Port au Prince, the capital of Haiti, and ruptured in the 2010 M 7.0 earthquake (shown by a star). Calais et al. (2010) give a more detailed description of the tectonic setting and earthquake history.

never be known, but in any case it was a devastating event for this poor, small country.

The earthquake occurred on or near the Enriquillo-Plantain Garden Fault zone, the southernmost of two major faults forming the boundary between the North American and Caribbean Plates (Figure 3.4). The region near this fault has been the locus of major earthquakes in the past, and it was clearly identified by seismologists as a zone of major risk. Documentation in the scientific literature of major earthquakes striking southern Hispaniola goes back nearly 100 years before the 2010 event and describes large earthquakes in 1751 and 1770. I made the first high precision GPS measurements of crustal strain in the region in the 1990s, and showed that the rate of strain accumulation on this fault was significant, a sure sign of future hazard. Dr. Eric Calais, then a professor at Purdue University and now at École Normale Supérieure in Paris, greatly refined these measurements, generated quantitative seismic hazard estimates, and actually met with government officials several years prior to the earthquake to warn them of the hazard. In the event, the various scientific studies and warnings proved to be about as useful in Haiti as they were at Fukushima in Japan.

Of course, unlike Japan, Haiti is a poor country and is unable to afford major infrastructure investments. But it does receive significant foreign aid; perhaps foreign donors could have been more diligent in requiring earthquake-proof construction? In fact, the dry, technical writing style of most of the scientific publications would have made it difficult for a donor or planner to understand the risk, let alone demand expensive upgrades. Let's look at the relevant summary sentence from my own scientific publication on the topic in 1998:

> The relatively high plate motion rate and fault slip rates suggested by our study, combined with evidence for strain accumulation and historical seismicity, imply that seismic risk in the region may be higher than previously estimated based on low plate rate/low fault slip rate models and the relatively low rate of seismicity over the last century. (Dixon et al., 1998)

While correct in strictly scientific terms, this statement gives no sense of urgency to the problem. Note also use of words such as "relatively," "suggested," "imply," and "may," a turgid writing style beloved by reviewers and editors that tends to blunt the impact of any conclusions. This style is typical of scientific publications and is part of the communication problem that is one of this book's themes.

After the earthquake, there was some discussion of moving the capital to a safer location, farther from the plate boundary fault. It is not usually feasible to move an entire city to a safer location, but the period immediately after a major disaster provides a unique opportunity. If significant infrastructure is destroyed and has to be rebuilt anyway, the marginal cost of building in a new, safer location may be low enough to make it worthwhile. Also, there is a brief window after a major disaster when people are prepared to discuss such changes. The 2010 Port au Prince earthquake destroyed so much infrastructure that such an opportunity existed, at least for a brief period. In the end, this did not prove feasible, but there was serious discussion of decentralizing government functions and critical infrastructure so that everything was not located in the same, vulnerable zone.

There was also lots of discussion immediately after the 2010 Port au Prince earthquake about the need to rebuild smarter, i.e. to a higher standard of quality, in conformity with internationally accepted earthquake standards. One challenge for Haiti is that the country lies in the tropical hurricane belt. Buildings here also need to be able to withstand hurricanes, which to some extent is at cross-purposes with earthquake requirements. For example, one way to engineer earthquake resistance is to have strong, laterally braced walls and a light roof. However, hurricane resistance usually implies a heavy roof that is able to withstand high winds. The 2010 Port au Prince earthquake is stimulating thinking on new designs that are not only resistant to both hurricanes and earthquakes, but can also be constructed at low cost.

Rebuilding in Haiti is now in progress, but it may be another a decade or more before we know how much the earthquake-engineering community has been able to influence new construction in Haiti and how much change the people of Haiti will be able to impose upon themselves.

## COST VERSUS BENEFIT

Dr. Seth Stein, a professor at Northwestern University who studies earthquake hazard and cost-benefit for earthquake-safe construction, has said, "Hazard mitigation is an investment today that reduces future losses." But how do we know how much to invest? If resources were unlimited, we could just build everything to extremely high standards, ensuring a high level of safety regardless of what happens. Since resources are limited, we need to be cleverer. In determining cost versus benefit, not only do we have to consider the size of today's investment against the anticipated future cost, but we also have to estimate when the future disaster is likely to occur, information that is usually poorly known. Stein and his father, Dr. Jerome Stein, a professor of applied mathematics at Brown University, have published a book that addresses the costs and benefits of natural hazard mitigation and describes a quantitative approach to the problem.

Getting the hazard estimate right is the obvious first step in the calculation. What is the probability of a magnitude 8 earthquake, or a Category 5 hurricane, hitting a particular location in a given period of time (for example in the next 50 years, the economic lifetime of many buildings)? These are statistical questions, and both the magnitude of the event and the time span to be considered are critical to the calculation. The likelihood of a major earthquake or hurricane occurring in a given location may approach certainty over the next 5,000 years, but the chances of getting hit in the next 50 years might be quite small.

The relevant time span for these calculations can differ depending on the structure and its projected usage. A parking lot is likely to be replaced with higher value construction long before the end of its physical lifetime, especially in rapidly growing urban areas. Thirty years is a long time in the parking lot business. While a given nuclear plant might only have an operating license for 30 years, when it comes time to upgrade, it may be difficult to get approval for an entirely new site. It would also be difficult to sell the site to new tenants, and it's expensive to demolish and decontaminate buildings such as nuclear containment vessels. An electric utility is more likely to upgrade existing facilities, leaving the contaminated buildings in place. Most nuclear power plants also act as storage facilities for high-level nuclear waste, at least temporarily. In the US, they are likely to retain this dubious distinction indefinitely. The reason is that, while the Yucca Mountain repository in Nevada was designed for long-term, high-level waste storage, for political reasons this facility has never opened, forcing power companies to retain their own nuclear waste on site. For all these reasons, once a nuclear power plant is built, that location is likely to retain nuclear facilities for a long time, well beyond the plant's original operating license. When calculating hazard for nuclear power plants, it would be prudent to use time spans that are much longer than the approximately 30-year design life of a particular facility.

Whenever we discuss risk, we also need to define what level of risk we are willing to bear. For example, if there is a one in a thousand chance ($P = 10^{-3}$) of a devastating earthquake at a given location in the next 50 years, is that an acceptable level of risk? Probably yes if we are building a parking lot; probably no in the case of a nuclear power plant. The US Nuclear Regulatory Agency defines an acceptable level of risk for loss of containment at a nuclear power plant as one chance in ten million per year ($P = 10^{-7}$ per year).

Some of these issues are relevant to the current debate about building codes in Memphis, Tennessee. Memphis is located close to the small town of New Madrid, which experienced a series of large earthquakes in 1811–1812. Early studies of these events suggested that they were quite large, of order M~8 (Johnston, 1996) nearly as big as the great San Francisco earthquake. If correct, and if such an event repeated today, it would be devastating, since most buildings in Memphis are not constructed in the same earthquake-resistant manner as buildings in California. Other estimates give somewhat lower magnitudes, closer to M~7 (Hough et al., 2000). Earthquake hazard maps published by the USGS show hazards in Memphis to be nearly as a large as those in California. Some geologists advocate strengthening of building codes in Memphis, because of the possibility of a repeat of the 1811–1812 events; others think that assignment of earthquake risk in Memphis at the same level of California makes little sense. Enforcing a strict building code on new construction and retrofitting existing buildings would cost billions of dollars. Given this uncertainty, the question of cost versus benefit is even more difficult to answer. Stein (2010) summarizes the scientific debate on the New Madrid fault zone.

Scientific debate and lack of consensus among scientists sometimes gets in the way of prudent action. One approach to deal with lack of expert consensus would be to implement only the lower cost aspects of mitigation until there is a higher level of consensus. In the case of New Madrid and Memphis, it would seem prudent to design new construction to a level appropriate for M 7 events, until such time as there is scientific consensus on whether M 8 or higher earthquakes are likely.

Forensic analyses of many disasters suggest that relatively inexpensive safety investments and basic common sense can prevent or greatly reduce the worst effects. In the case of New York's Triangle Shirtwaist Factory fire, simply training floor managers to let people out of the building in the event of a fire would have greatly reduced the loss of life. You might think that such training should not be required – it's too obvious. Unfortunately, you would be wrong: Fatalities from fires where people had difficulty leaving the building continue to occur, often with large death tolls. Boston's Cocoanut Grove nightclub caught fire in 1942, killing nearly 500 people. The owner had blocked exits so people couldn't leave without paying their bills. A dance club in Brazil experienced a fire in 2013 that resulted in more than 240 deaths for essentially the same reason. Bouncers guarding the exits thought customers were trying to skip out of paying their bills, and refused to let them leave. The fire was started by pyrotechnic devices (fireworks) used inside the club as part of the show. Since 2003 there have been at least six nightclub fires with significant fatalities that involved blocked exits. Fireworks were implicated in most of these. Affected countries include Argentina, Brazil, China, Romania, Thailand, and the US. Do we really need laws that explicitly state that fireworks should not be used indoors or that in the event of a fire, people should be allowed to leave?

Failure to train managers to think critically in emergency situations or take simple safety measures can have disastrous effects on a business. Students of free market capitalism often use the term "creative destruction" to describe the tumultuous but generally beneficial process of the rise and fall of companies as markets and technologies change. The concept was first articulated by Austrian economist Joseph Schumpeter in *Capitalism, Socialism, and Democracy*, published in 1942, although elements of the idea were originally proposed by Karl Marx. A good example is the change in the music industry in the last decade, as major record labels, who once sold most music via vinyl records and compact discs, now face stiff competition from companies such as Apple who sell music over the Internet. But companies and industries can fail for more reasons than technological change. Failure to manage risk associated with natural events and disasters can also lead to business failure, a process we might call "non-creative destruction," since viewed in hindsight, the failures seem easily avoidable. This book unfortunately provides numerous examples of non-creative destruction of otherwise promising facilities and businesses through poor location and design, inadequate investment in basic safety and hazard resilience, hiring and promotion of poor managers, and a short-term focus on profits at the expense of longer-term considerations of public health and welfare.

Another aspect of cost-benefit analysis related to disaster-resistant construction and management that is rarely discussed, at least in public, is the value placed on a human life. On the benefit side of the equation, a major goal of good planning, design, and construction is the preservation of human life. Assuming we value life highly, it would seem to make sense to put a lot of money into safe construction. But how much? If a community puts lots of money into disaster-resistant infrastructure, there will be less funding available for health care or salaries for teachers or firemen. When we make such calculations, what value should we assume for a human life? And are there different values for different people? Should schools be made safer than other buildings because they house children with their entire lives ahead of them? At first glance this seems to make sense. But does this imply that retirement homes should be made less safe, on the grounds that the residents have less time to live? Are we prepared to put grandma in an unsafe building? These are challenging questions, with a host of ethical, cultural, and economic implications that go far beyond the science.

The medical community has had to wrestle with such issues when dealing with organ transplants. For certain organs (such as livers), demand greatly outweighs supply, so the medical transplant community has to decide who will get the scarce organ, in effect deciding who will live and who will die. Children do not automatically come first in these decisions; patient health and long-term survivability also come into play.

The legal system has also had to wrestle with these questions, usually for cases involving liability. New York courts decided in 1914 that an award of $75 to each victim in the Triangle Shirtwaist Factory fire was satisfactory (Box 2.1). Awards in New York have become more generous with time. Families of the 9/11 victims received an average of $2 million in compensation, using a formula that included not only age at the time of death, but also estimated future earnings power; awards ranged from $250,000 to more than $7 million, with higher awards tending to go to higher income individuals. If we were to follow this formula in deciding how safe to make buildings, we would engineer them to be extremely safe if they are to house highly paid CEOs, financial consultants, and professional sports players, but we would build more cheaply for people with lower incomes.

Beyond the obvious practical problems (Would you ask to see people's tax returns before they entered a building?), most people would strenuously object to income-based building codes on moral and ethical grounds. But the principle of income-based standards seems strongly entrenched, at least in the US legal system, where liability awards are routinely based on lost earnings potential. Most countries implicitly accept the concept of income-based standards for privately financed residences. Low- and middle-income housing generally meets some minimum legal standard, established by national, state, or local government, but high-income housing is usually built to higher standards. When tornadoes strike the US, it is not a coincidence that the scenes of destruction that play out on television news are usually filmed in trailer parks and other low-income housing. Mobile homes, often the shelter of last resort for retirees and other fixed-income groups, are especially vulnerable to wind damage. (It is also true that, given unequal income distribution, many more people live in trailer parks and other low-income housing compared to high-end housing.) Simple tie-downs would increase the survivability of mobile homes, at a cost of a few thousand dollars per unit or less. This low-cost remedy would reduce risk for many people in the tornado belt, and would be a good investment for the nation. Florida, with nearly a million mobile home owners, many of them fixed-income seniors, has a program that subsidizes trailer tie-downs and other mobile home safety measures. This makes economic sense and at least partly addresses some of the ethical issues noted above.

In the next chapter, I'll apply the lessons of this chapter and the scientific background on subduction zone earthquakes and tsunamis described in Chapter 2 to the 2011 Fukushima disaster, and then I'll consider to what extent that disaster could have been minimized.

# 4 Japanese Earthquakes and Nuclear Power Plant Failures

When I was a graduate student in California in the 1970s, the scientific study of hazard from active faults was in its infancy. Geologists were being hired by geotechnical and environmental consulting firms to perform risk assessments for large projects. Many of these projects had already been started, or were even completed, but new laws required seismic hazard assessments to be performed anyway to identify nearby seismogenic faults (faults capable of producing damaging earthquakes), and upgrade infrastructure if necessary. A joke circulating at the time was that if you wanted to find an active fault in California, you just had to look where the nuclear power plants were located. Obviously things are a bit more complicated, but it does illustrate the point that some nuclear power plants are in locations now recognized as unsafe.

In this chapter, I'll review causes and consequences of the earthquake and tsunami that struck northern Japan in 2011 and destroyed the Fukushima Daiichi nuclear power plant. I'll use concepts developed in the first three chapters, including subduction zones (Figure 2.4), strain accumulation, release and tsunami generation (Box 2.4), and aliasing and clipping (Box 3.1) to assess whether the plant was situated in a safe location, whether it was designed well, whether information available before the accident was properly utilized, how much we actually know about the scale of the disaster, and how we know it. I'll also discuss other problems that contributed to the disaster, problems that are unfortunately a common theme in many modern disasters. These problems include a lack of transparency that makes it difficult for the public, the media, or outside experts to evaluate decisions and risk trade-offs made by government officials and businesses; strong vested interests that put profits ahead of the larger public good; and cozy relationships between regulatory agencies and the industries they supposedly oversee.

The 2011 earthquake and tsunami at Fukushima in northern Japan were unavoidable, but most of Japan's well-designed, earthquake-resistant infrastructure performed well. The country's network of strong motion sensors gave residents in Tokyo a one-minute warning of imminent strong shaking, and the tsunami warning system gave coastal residents 10 to 30 minutes to evacuate to higher ground. In spite of these well-designed defenses, a cascading series of largely avoidable failures occurred at the Fukushima Daiichi nuclear complex.

Much of Japan lies in a subduction zone, where the Pacific Plate is pushed beneath the Eurasian and Phillipine plates (Figures 2.2–2.4). As we learned in previous chapters, this type of plate boundary produces Earth's largest earthquakes and most tsunamis. Since 1950, subduction zones have produced five extremely large earthquakes, with moment magnitudes (a measure of their total energy) of 9.0 or greater (Table 4.1). Three of these (Chile in 1960, Alaska in 1964, and Sumatra in 2004) were larger than the Japanese earthquake and also produced devastating tsunamis. In terms of energy release, the 2004 Sumatra earthquake released roughly 50 percent more energy than Japan's recent event, so things at Fukushima could have been worse.

Table 4.1 covers just over 100 years, a period during which we have had good information from seismometers, sensitive instruments that record ground motion from earthquake shaking. Several interesting points stand out. First, nine out of ten events occurred at subduction zones where oceanic crust is subducted (Figure 2.4). Tibet is the exception; it occurred in a continental collision zone (Figures 3.1 and 3.2). Second, most subduction zone events are in the Pacific "Ring of Fire" (Figure 2.3). The two Sumatra events are the exception. Third, there is some evidence of clustering: Two events in Alaska and two events in Sumatra occurred within a year of each other.

Table 4.1 *Ten largest earthquakes in the last 105 years*

| Location | Year | Magnitude |
|---|---|---|
| Chile | 1960 | 9.5 |
| Sumatra | 2004 | 9.2* |
| Alaska (Prince William Sound) | 1964 | 9.2 |
| Japan | 2011 | 9.0 |
| Kamchatka | 1952 | 9.0 |
| Chile | 2010 | 8.8 |
| Ecuador-Colombia | 1906 | 8.8 |
| Alaska (Rat Island) | 1965 | 8.7 |
| Sumatra (North) | 2005 | 8.6 |
| Tibet-Assam | 1950 | 8.6 |

* Average of a range of published values. USGS/NEIC magnitude is 9.1. Stein and Okal [2005] give the magnitude as 9.3. The range reflects in part the complexity of the rupture process for this large event.
*Source:* US Geological Survey/National Earthquake Information Center

This could be a coincidence, but could also reflect the fact that some large earthquakes "load" an adjacent segment of the subduction zone, making it likely that the next earthquake will occur sooner rather than later (see Chapter 5). Fourth, for these extremely large events, such as Sumatra in 2004, it's hard to know the exact size of the earthquake. The ruptures are so long, in both space and time, that different approaches used to calculate earthquake size can shape the result. Finally, the 2011 Japan event was the first "top ten" earthquake to hit the island nation since this type of detailed record keeping became possible.

While the 2004 Sumatra event may have been a bigger earthquake, the 2011 Tohoku earthquake was, nevertheless, big enough to cause serious devastation. Sites on land jumped eastward by up to 5 meters, as measured by Japan's network of high precision GPS stations. Sites on the sea floor, closer to the surface trace of the plate boundary (the Japan Trench) moved up to 50 meters, with correspondingly large vertical motions of the sea floor, ultimately leading to the large tsunami (Box 2.4).

Despite these huge motions and attendant severe shaking, the nuclear plant at Fukushima came through the earthquake with flying colors. The reactor "scrammed" as it was designed to: Control rods were lowered into the nuclear fuel, absorbing neutrons and abruptly stopping the nuclear chain reaction. Though radioactive decay continued, generating heat, backup generators kicked in to keep cooling waters circulating around the fuel, preventing overheating. When an earthquake affects both the plant itself and the nearby electrical grid, redundant sources of onsite backup power are critical.

The real problem came 30 minutes later, when the tsunami hit. Because backup generators were not elevated or watertight, they were immediately knocked out by flood waters. Flooding also shorted out backup control functions. While the plant had been designed to withstand severe earthquake shaking, little thought had been given to how it would handle a major tsunami.

This is bizarre, because the critical importance of backup power for coolant pumps is well known in nuclear power plant design and operation. Designers generally factor in large safety cushions. In fact, the plant is located at the site of a natural coastal ridge, high above sea level. Oddly, engineers went to great trouble and expense during construction to remove the coastal ridge so the entire plant could be located on a single flat surface near sea level. Plant designers were apparently unable to conceive of the possibility of a tsunami and the advantages of a multilevel plant with seawater intakes at low elevation and backup power and control at higher elevation (note to future nuclear plant designers – please think in 3-D!).

The worst was yet to come. Mobile generators that were rushed to the scene a few hours later apparently had the wrong kind of connectors and could not interface with the cooling pumps. Plant managers were woefully unprepared for an emergency.

The various missteps set the stage for a nuclear plant operator's worst nightmare – a core meltdown. If nuclear fuel cannot be cooled, it begins to overheat and will melt through the structure holding it and, eventually, through the floor of the containment vessel, potentially reaching the water table and exploding, leading to widespread contamination. The phrase "core meltdown" was studiously avoided by executives at TEPCO (Tokyo Electric Power Company, the owner and operator of the plant), but almost certainly, partial meltdowns occurred at several of the Fukushima reactors. Hydrogen gas, which is quite explosive (20th-century-history buffs may recall the Hindenburg, Box 4.1), began reaching high concentrations in several of the containment vessels. Modern plant designs usually contain

### BOX 4.1 **Hydrogen and the Hindenburg**

The Hindenburg was a German dirigible (airship) made lighter than air by virtue of a large volume of hydrogen gas. The same type of explosion that crippled the Fukushima Daiichi reactors destroyed the Hindenburg on its maiden voyage from Frankfurt, Germany, to Lakehurst, New Jersey, in 1937. Prior to the explosion, airship travel was considered the up-and-coming way to cross the Atlantic, superseding slower surface ships. Although Alcock and Brown's 1919 Atlantic crossing and Lindbergh's 1927 crossing with airplanes had already occurred, airplanes were not yet capable of carrying large numbers of passengers across the Atlantic. The return trip on the Hindenburg was sold out, and a new type of travel industry seemed possible. Regularly scheduled trans-Atlantic flights between Germany and Brazil in the Graf Zeppelin, another dirigible, had occurred since 1931, but the Hindenburg was a larger and more comfortable ship, ready to crack the all-important US market. However, the loss of life and catastrophic destruction of the Hindenburg, which was caught on film as well as live radio, effectively ended the airship era. Regularly scheduled passenger service with airplanes would not start for another two years, with Pan Am's inaugural run between New York and France using a Flying Boat. Trans-Atlantic jet service would not begin until 1958.

BOX 4.1 **(cont.)**

passive catalytic systems that automatically react hydrogen with oxygen to form water, keeping hydrogen concentrations at safe (low) levels. These systems either were not installed or did not operate properly at Fukushima. The net result of these design and management flaws was at least one and possibly three partial core meltdowns and a series of explosions that breached one or more containment vessels, resulting in massive releases of radioactive contaminants. The resulting radioactive cloud could be followed by radiation sensors as it circled the globe. Although people outside Japan were exposed to extremely small amounts of radiation (much less than a typical dental X-ray), it is probably fair to say that the event did not inspire much confidence in nuclear power, in TEPCO, or in Japan's regulatory agency. Nöggerath et al. (2011) give a complete description of various problems at the power plant.

### HYDROGEN: WHY IS IT PRODUCED IN NUCLEAR POWER PLANT ACCIDENTS? WHY IS IT DANGEROUS? WHAT'S IT GOT TO DO WITH THE HINDENBURG?

Nuclear fuel rods consist of uranium oxide pellets in zirconium tubes. If coolant stops circulating, the rods heat up, eventually cracking and releasing radioactive products. If the tubes continue to heat and reach a temperature of around 1200°C, the zirconium metal will react with steam to form an oxide of zirconium and hydrogen gas:

$$\mathrm{Zr} + \mathrm{H_2O}\,(\text{steam}) \longrightarrow \mathrm{ZrO} + \mathrm{H_2}(\text{gas}) \tag{4.1}$$

Once produced, the hydrogen gas accumulates in the containment vessel or outer building. Hydrogen gas is extremely reactive, and becomes explosive at concentrations higher than a few percent, recombining with oxygen to form two water molecules in a violent reaction:

$$2\mathrm{H_2} + \mathrm{O_2} \longrightarrow 2\mathrm{H_2O} \tag{4.2}$$

The resulting explosion can be strong enough to rupture even massively reinforced concrete containment vessels, which normally sequester the radiation, as appears to have happened at Fukushima.

TEPCO's disastrous design and management at Fukushima has the potential to end the era of nuclear power, just as the Hindenburg's demise ended a promising airship era. Nuclear plant construction and planning around the world has slowed or stopped since the accident, and opponents of nuclear power have been emboldened. Germany decided to phase out its nuclear power plants by 2022 as a direct result of the accident. While there are obvious differences between the catastrophic explosions of the Hindenburg and at Fukushima, engineers, managers, and marketing experts would do well to relearn the lesson that putting large amounts of hydrogen in one place and letting it explode can not only be fatal to people, but also be fatal to a promising industry.

## HOW BAD WAS FUKUSHIMA?

TEPCO and Japanese authorities initially assured the Japanese population that the radiation produced by the accident was much lower than the 1986 Chernobyl incident (a massive nuclear accident in the former Soviet Union that dispersed radiation around the globe) and, hence, posed minimal health risk. For example, on April 11, 2011, Japan's nuclear safety agency, relying in part on TEPCO sources, stated that the nuclear accident had released about 10 percent of the radiation released by Chernobyl. How this claim could be made at the time is not clear to me, since the amount of radiation released in the accident had not yet been measured in any reliable way. It is actually quite difficult to estimate total radioactivity released in such an event. There are three main problems – when, what, and where.

**When:** The best time to make such measurements is during and immediately after the accident. The company's own automated, onsite measurement systems were best positioned for this, but most of these were damaged or destroyed in the tsunami. In addition, the most sensitive instruments operating near the site, designed to measure small amounts of radiation, went off scale (see Box 3.1). Instruments with higher dynamic range, designed to deal with much higher levels of radioactivity, will not necessarily be located in exactly the right places (see "Where," below). Thus, the maximum levels of radiation released during the accident go unrecorded, and must be estimated later through indirect techniques. Even if good near-field measurements (close to the accident site) are available, a private company may be reluctant to release the most damaging data due to concerns regarding future liability; there is a strong financial incentive to low-ball the estimate. Measurements by independent scientists are important in situations like this, but it takes time for scientists to organize a big measurement program, make travel arrangements, and obtain the necessary permissions and funding.

**What:** Hundreds of different radioactive contaminants were released in the accident and contribute to harmful radioactivity. Measuring the various contaminants may require different procedures and different instruments. There are also huge ranges in the amount of radioactivity depending on location and timing. The most precise measurement systems may only focus on one or at most a few radioactive contaminants and may not work at high levels of radioactivity. Measurements reported immediately after a major accident are not necessarily the most important in terms of health effects, but rather the ones that are easiest to make on short notice and on a limited budget (Box 4.2)

### BOX 4.2 **Measuring Fukushima's radiation: The role of cesium**

Measuring radiation releases at Fukushima has been challenging. One of the questions is what to measure – there are hundreds, possibly thousands, of radioactive byproducts at the accident site, and some instruments only focus on one or a few species. Many studies rely on a radioactive version of cesium (Cs), $^{137}$Cs (read as Cesium-137) with a half-life of 30 years. When a uranium atom splits (fissions) in a nuclear reactor, $^{137}$Cs is a common byproduct, making it an excellent tracer for radioactivity at nuclear power plants.

Another reason that Cs is an excellent tracer for nuclear accidents: It is not produced naturally. Any $^{137}$Cs in the environment today was produced in 1942 or later by an artificial process, such as a nuclear bomb explosion or a nuclear reactor accident. (December 1942 is the date when the first artificial nuclear reactor, part of the Manhattan Project, began operating in Chicago.) Most $^{137}$Cs began to show up in the environment after 1954, when nuclear weapons testing started in earnest. Shortly after the accident at Fukushima, levels of $^{137}$Cs were many orders of magnitude higher than background levels measured before the

BOX 4.2 **(cont.)**

accident, because $^{137}Cs$ released during the accident resulted in concentrations much higher than residual levels left over from decades earlier weapons testing in the western Pacific.

This history and its 30-year half-life make $^{137}Cs$ an excellent tracer for all sorts of things beyond nuclear accidents, including atmospheric and ocean circulation, sediment, snow and ice accumulation rates, and groundwater movements. The range of applications also explains why there are lots of scientists and graduate students able to measure and interpret radiation data from Fukushima. Scientists are not just sitting around waiting for a nuclear accident to happen; they can use artificially produced tracers like $^{137}Cs$ to study any number of important natural processes. An ironic side effect of the Fukushima accident is that it will be an excellent "point source" (in both space and time) for future tracer studies of many natural processes. It has already been used to study snow accumulation rates in the Arctic. In the future, it will probably be used as a marker in Greenland's ice cores, reliably marking 2011 as the year when radioactive fallout from the accident first began to accumulate in that continent's snow and ice layers.

**Where:** There are several contamination sources and pathways that need to be considered. In terms of sources, three different reactors were operating at the time of the accident. There were also several pools for spent fuel rods; these can experience damage during the earthquake and tsunami, releasing radiation to the environment immediately, or in the absence of active cooling, they can overheat weeks later and then release radiation. Contamination pathways include atmospheric fallout over land, atmospheric fallout over the ocean, and direct discharges of contaminated water into the ocean by several routes. These include cooling water sprayed over the reactors that streamed directly into the ocean, or leaked from temporary impoundments into the ocean, or leaked into the groundwater and later into the ocean. There are also dozens of tanks holding radioactive wastewater. Finally, damage to the reactor cores' containment vessels may have included damage to their floors. If so, this also provides a pathway between radioactive products and the external environment, through the groundwater table and subsequent movement into the ocean. Most of these contamination sources did not have measurement systems nearby, where signals would be strongest. While the world has a network of ground-based sensors maintained by CTBTO (Comprehensive Nuclear Test Ban Treaty Organization) for measurement of atmospheric contamination, there is no equivalent for the ocean, where most of Fukushima's radiation actually wound up. By the time a rigorous measurement program could be initiated, radioactivity from the accident was already widely dispersed across a broad expanse of ocean, contaminating marine life hundreds or even thousands of kilometers from the site and making detailed characterization extremely difficult.

For several weeks after the accident, the uncertainty concerning its magnitude and the lack of transparency on the part of TEPCO officials about radiation levels led to considerable public unease, especially for those living within a few hundred kilometers of the plant. Joichi Ito, currently head of the Media Lab at the Massachusetts Institute of Technology, was so frustrated with the lack of information (his family was living in Japan at the time) that he helped set up an open-source monitoring network (safecast.org) that makes information on radiation levels publicly available. His TED talk detailing this experience (http://www.ted.com/talks/joi_ito_want_to_innovate_become_a_now_ist) makes for fascinating viewing.

In the years since the disaster, measurements and models by the Japanese and international scientific community have clarified the amount of radiation released during and after the accident, and the various pathways by which this radiation has entered and continues to move through the global environment. Box 4.2 outlines one of the ways this is done, using a radioactive version of the element cesium. Published estimates for the total radiation released at Fukushima are in the range of 25 percent of Chernobyl's levels (the most recent estimate as of this writing and probably the most accurate) to as high as 60 percent (Box 4.4); the wide range reflects the difficulties described above in making such estimates.

$^{137}Cs$, a radioactive version of cesium, has many industrial applications, meaning that instruments that measure it are available at relatively low cost, a big plus for scientists working on shoestring budgets. $^{137}Cs$ is even used to test wine (Box 4.3).

BOX 4.3 **Cesium and wine**

Most northern hemisphere wine produced after 1954 has measurable amounts of $^{137}Cs$. Southern hemisphere wines have less, since atomic weapons testing mainly occurred in the northern hemisphere. Thanks to Fukushima, northern hemisphere wines made from grapes harvested after 2011 will likely have a little more $^{137}Cs$ compared to earlier vintages.

Wealthy collectors will pay tens of thousands of dollars or more for old wine. Huge sums have been paid for wine that is supposedly from the cellar of Thomas Jefferson, third president of the US and a well-known wine connoisseur. Such wine would be more than 200 years old. Not surprisingly, there is a thriving market in fraudulently labeled old wines. $^{137}Cs$ can be used to verify whether a wine was produced before or after 1954. In the famous case of the Jefferson wine, it turns out that the wine is actually much younger. Patrick Keefe, in an article published in *The New Yorker* (2007), and Benjamin Wallace, in his book *The Billionaire's Vinegar* (2008), give fascinating accounts of this arcane world at the intersection of wine, wealth, ego, fraud, foolish investment, and radiochemistry.

Even the lowest published estimates of radiation released by the Fukushima accident (Box 4.4) are more than double the 10 percent of Chernobyl's levels initially reported by TEPCO and government officials. This raises the issue of transparency by the company, a recurring topic in this book related to the theme of communication. Company and government officials either were not truthful with the public or were ignorant of the hazards posed by the accident. Neither is

### BOX 4.4 **Comparisons to Chernobyl**

Radiation is often measured in units of Becquerels (Bq), where one Bq = one nuclear disintegration per second. Any time a radioactive particle disintegrates, it gives off potentially harmful radiation in the form of alpha or beta particles or gamma rays; Becquerels account for all such disintegrations without distinguishing between them ($^{137}$Cs gives off both beta particles and gamma rays). A banana gives off about 15 Bq of radiation, mainly from decay of $^{40}$K, a naturally occurring radioactive form of potassium. At its peak, Chernobyl gave off about 85 PBq of radiation just from $^{137}$Cs, where P stands for peta, meaning $10^{15}$ (10 with 15 zeros). As a colleague of mine pointed out, that's a lot of bananas.

Stohl and others (2011) estimate Fukushima's atmospheric release of $^{137}$Cs at about 37 PBq. Rypina et al. (2013) estimate direct releases to the ocean of 16 PBq. Summing these two estimates gives a total of about 53 Pbq, or about 60 percent of Chernobyl's levels of $^{137}$Cs. More recent published estimates are lower (see References and Further Reading), in the range of 25–35 percent of Chernobyl's levels of $^{137}$Cs. The range in published values (25–60 percent, more than a factor of two) is an indication of the difficulty of assessing the total radiation released by this accident, especially during and immediately after the event. Note that the lowest published estimate (25 percent of Chernobyl) is more than double TEPCO's initial estimate (10 percent of Chernobyl).

a comforting thought. Reports in October 2012 by Reuters, the *New York Times*, and other news organizations suggest that TEPCO officials sought to limit future liability by downplaying the size of the accident. In response to pubic pressure, the government eventually decided to shut down the plant and all other nuclear power plants in Japan. While some plants are currently being reopened, the damage to TEPCO's credibility, Japan's economy, and the global nuclear industry has been severe.

TEPCO uses the term "cold shutdown" to describe the current state of the damaged reactors, a term meant to reassure the public. However, as of late 2012, contamination continued to flow into the ocean (Buesseler, 2012), and presumably continues today. The contamination pathways include groundwater that is receiving contamination from one or more damaged reactor cores, and waste ponds and tanks on site that are leaking. For a long time, TEPCO denied that there was any leakage into the ocean, despite overwhelming scientific evidence. A French report in 2012 stated that the flux of radioactive leakage into the ocean was 20 times higher than TEPCO reports. Moreover, the company failed to take even basic precautionary measures to assess the problem. An August 24, 2013 article in the *Economist* noted that two and a half years after the earthquake, tanks holding radioactive wastewater still lacked basic systems to monitor leakage. Instead, employees climbed up the side of the tanks, peered in, and tried to memorize the water level for comparison to a future inspection. Continued groundwater leakage from the site means that estimates of total radiation released by the accident are not well known. Rather than cold shutdown, it would be more accurate to characterize the reactors as being in a state of "warm leakage."

Dumping of radioactive waste into the sea is a violation of international law. It is not clear if TEPCO's continued release of radioactively contaminated wastewater into the ocean constitutes a violation of this accord. This may depend on whether TEPCO officials can convince investigators from the IAEA (International

Atomic Energy Agency) that they have not done so on purpose, but are simply unable to manage their wastewater issues as a result of prior negligence. Again, neither possibility is a comforting thought. I am puzzled that the IAEA has not pursued this apparent violation of international law more vigorously.

Long-term studies of Chernobyl suggest that it has been responsible for approximately 5,000 additional cases of cancer, mainly thyroid cancer among people exposed to radiation at a young age (UNSCEAR, 2008). Cancer rates in Japan due to the accident will be much less, since most of the original radiation cloud blew out (or was flushed out) to sea, limiting exposure to individuals near the plant. It will be many years before comparable statistics on cancer levels and fatalities are available for the Fukushima disaster. A much larger marine area adjacent to the plant is also contaminated and will be closed to fishing for many years. Tens of thousands of people were displaced from their homes and farms. The long-term costs of the accident are not yet known, but will probably exceed $200 billion. In a two-year business plan released in November 2012, TEPCO estimated costs at $137 billion. Dr. Kazumasa Iwata with the Japan Center for Economic Research estimates that costs will be between $70 billion and $250 billion, with major costs covering purchase of contaminated land and scrapping of damaged reactors. A World Bank estimate put total costs at $235 billion. This would make it the costliest natural disaster in history, exceeding by more than a factor of two costs from Hurricane Katrina, the previous "winner" (see Chapter 7).

Of the many unfortunate consequences of this accident, one of the longest lasting may be the public's loss of confidence in official reporting on technical issues, especially regarding nuclear power. This is at least in part a consequence of TEPCO's initial reports on the incident, which lacked credibility and transparency. The resulting information vacuum fueled scare-mongering and conspiracy theories. Internet searches in 2015 reveal dozens of dubious articles claiming that various disasters were related to Fukushima. As bad as

Fukushima was, it is important to keep perspective – far more people die in car accidents every year than will die from Fukushima's radiation releases. Chapter 6 discusses the concept of relative risk and compares nuclear power with other sources of energy.

## WHAT WENT WRONG?

The design standard used at Fukushima Daiichi relied on an outmoded seismic model, that of the characteristic earthquake. This model, in vogue when the plant was originally designed 50 years ago, holds that earthquakes that have occurred in a given location over the last few hundred years (the time period where reasonably accurate earthquake-size estimates can be made) will be repeated in the future with similar characteristics and, hence, can tell us something useful about earthquake hazard in the future. Using this standard, northeast Japan was unlikely to face anything larger than a magnitude ~8 earthquake, since nothing in the recent past had exceeded that size. Such an event, though significant, makes a devastating tsunami less likely. There are two problems with this assumption.

First, it turns out that tsunami size is not directly related to earthquake size. You can have a big tsunami even with a magnitude 8 or smaller earthquake. Second, the historical record is short compared to the time scale for earthquake recurrence. We'd have to go back several thousand years at a given location if we wanted to use the earthquake record. Since the seismometer, the instrument that measures ground shaking due to earthquakes, was only invented about 120 years ago, it would appear that such data are unavailable. In most places, the historical earthquake record is too short to be representative (Box 3.1).

How do geologists put together a longer record of earthquakes and avoid this problem? They use Steno's principle of superposition (younger stuff on top; see Box 1.1). Figure 4.1 shows an example specific to earthquakes.

The example in Figure 4.1 is pretty simple, because only one earthquake has disrupted the sedimentary layers. In reality, several

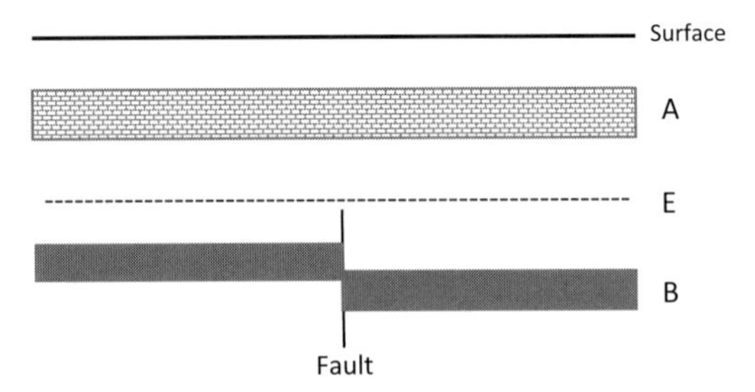


FIGURE 4.1 Hypothetical and simplified cross section of the vertical wall of a trench dug for paleoseismology research. A fault (a fracture in the earth that accommodates differential motion, usually via earthquakes) has displaced an older sedimentary layer (B) but not a younger layer (A). The exact time of the earthquake that produced the fault offset is not known, but is probably close to the time of deposition of a thin layer (E), called the "event horizon." Ideally the age of E can be determined. If not, charcoal in units B (5,000 years old) and A (3,000 years old) is datable by radiocarbon techniques. Ignoring uncertainty in the radiocarbon dates, this suggests that the earthquake occurred 4,000 plus or minus 1,000 years ago.

events may be exposed in a given trench, producing a complex web of geometric relationships that takes considerable skill to decipher. Despite the complexity, such locations are invaluable for defining the recurrence history of multiple earthquakes and tsunamis. McCalpin (2009) has produced an excellent textbook on the topic.

Trenching of active faults became common in the 1970s, especially in California, as new regulations required nuclear power plants and other critical facilities to determine whether they were located on or near active faults. Geologist Kerry Sieh, then a graduate student at Stanford University and later a professor at Caltech, was studying the earthquake history of the San Andreas Fault in Southern California. His detailed trenching studies at Pallett Creek, published in 1978, became a classic paper showing how geometric relationships and radiocarbon dating could elucidate the earthquake history of a fault. This work conclusively demonstrated that big earthquakes happen frequently (roughly every 200 years) on the San Andreas Fault and established trenching as a key tool in paleoseismic studies. Similar work has now been conducted worldwide on most active faults, especially in the vicinity of critical facilities such as nuclear power plants. In Japan, scientists broadened these studies to look at both earthquake and tsunami deposits, as described later in this chapter.

Cracks in the characteristic earthquake model first started to appear in the mid 1990s. A long-running experiment in Parkfield, California, was predicated on this model. M ~6 earthquakes had occurred in Parkfield on average every 22 years since the early part of the 20th Century, with the last one occurring in 1966. The area appeared ripe for another similar earthquake beginning in the mid or late 1980s, with the model predicting a very high probability of an event before 1993. In anticipation, a large amount of geophysical monitoring equipment was installed. However, nothing happened until 2004 – well after the "prediction window" had closed – when a magnitude 6 earthquake occurred close to the predicted area. It was the right size, but the wrong time. Could the model also be flawed in terms of size prediction?

The characteristic earthquake model had been out of favor since the mid and late 1990s, but the giant Sumatra earthquake in 2004 was its death knell. A corollary of the model was that the geologic characteristics of each subduction zone could be used to predict the maximum size of earthquakes that could happen there. No earthquake of this size had been forecast for Sumatra, and the region did not meet the model's criteria for giant earthquakes (fast subduction of geologically young ocean crust). An uncomfortable truth was recognized by many scientists after the 2004 event: The characteristic earthquake model could predict neither the timing nor the size of future events. Given current knowledge, we had to admit that an event the size of the 2004 Sumatra earthquake could happen in any subduction zone.

In a paper published in January 2007, Drs. Seth Stein and Emile Okal, professors at Northwestern University, commented on the different magnitudes of the biggest earthquakes in different subduction zones and stated, "We suspect that much of the apparent differences between subduction zones ... reflect the short earthquake history sampled." In other words, our record is too short, and we haven't yet seen the biggest earthquakes and tsunamis that can happen on many subduction zones. Nuclear power plants and other key infrastructure situated in these zones need to be engineered accordingly.

Dr. Robert McCaffrey, then a seismologist at Rensselaer Polytechnic Institute in New York, wrote in 2007 in the journal *Science*: "For policy purposes, one lesson we should take away from the Sumatra-Andaman earthquake is that every subduction zone is potentially locked, loaded, and dangerous ... We should, by sustained education, embed the lessons of 2004 in the cultural memories of all coastal communities." In a subsequent article in 2008, he suggested that every subduction zone should be considered capable of producing a magnitude 9 earthquake. Most geologists agreed. Unfortunately, this consensus was not enough for TEPCO.

More generally, for any type of plate boundary, if one wanted to be prudent, a good indicator of earthquake hazard would simply be the maximum known magnitude of earthquakes on that type of plate boundary, no matter where they occur on Earth. If you happen to live on or near that type of boundary, you should assume that a similar magnitude event is possible at your location, even if one has never occurred there before. In effect, this trades space for time. In other words, since we usually can't record all the hazardous events that could happen at a single location over thousands of years, the next best thing is to look around the world at similar places and try to understand the size and frequency of disasters of similar type. For Japan, that means looking at all subduction zones around the world and recognizing that magnitude 9 earthquakes were a possibility. Using this criterion, TEPCO's managers and engineers had at least four years (from 2007, when McCaffrey published his article in *Science*, to 2011) and perhaps seven years (from the 2004 earthquake in Sumatra) to raise their generators and beef up their backup systems. The IAEA specifically asked TEPCO to revaluate its design standard and re-engineer its cooling systems after the 2004 Sumatra earthquake; apparently TEPCO ignored the request (Connor, 2011). The accident response plan that was in place for the Fukushima plant at the time of the accident stated, "The possibility of a severe accident occurring is so small that from an engineering standpoint, it is practically unthinkable."

In the immediate aftermath of the disaster, TEPCO executives and some Japanese politicians portrayed the event as an act of God, an unpredictable disaster that no one could have foreseen (one of Talib's "black swans"). If true, this would limit the legal, economic, and ethical liabilities of TEPCO executives. Warnings were clearly available. But were they delivered clearly, and to the right people? As with the L'Aquila example in Italy, the issue comes down to communication.

In fact, it had been known for some time in Japan (much earlier than 2007) that both earthquake behavior and tsunami generation are far more complex than implied by simple characteristic earthquake models and that huge tsunamis had occurred in or near the Fukushima region in the past. Geologists know that the historical record is too short to be representative. In northern Japan, this knowledge is based on trenching, mapping, and similar studies that investigate earthquake and tsunami deposits in great detail. For example, a paper published in 2003 in the widely read journal *Nature* by Dr. Futoshi Nanayama of the Geological Survey of Japan, Dr. Brian Atwater of the US Geological Survey, and other scientists documented unusually large tsunami deposits in northern Japan, hinting that much bigger events were possible than indicated by the historical record. Here is part of the first paragraph of their report:

> Along the southern Kuril trench, which faces the Japanese island of Hokkaido . . . fast subduction has recurrently generated earthquakes with magnitudes of up to 8 over the past two centuries. These historical events, on rupture segments 100–200 km long, have been considered characteristic of Hokkaido's plate-boundary

> earthquakes. But here we use deposits of prehistoric tsunamis to infer the infrequent occurrence of larger earthquakes generated from longer ruptures.

Although this report focused on Hokkaido, in northernmost Japan (north of Fukushima), the geologic characteristics of the subduction zone there are similar to the situation offshore Fukushima. Moreover, there was abundant evidence that such large tsunamis were possible, even likely, at Fukushima long before the 2003 publication on past tsunami deposits in Hokkaido and long before the 2004 Sumatra earthquake changed scientific opinion on the characteristic earthquake model.

Historical records document a large tsunami at essentially the same location as the 2011 event, a little more than a hundred years earlier. In 1896, a large tsunami struck near Fukushima, related to the 1896 Sanriku earthquake. Although the earthquake was much smaller than the 2011 event, the tsunami was surprisingly similar, at least in terms of location and run-up distance and height (run-up height is the maximum elevation affected by the tsunami, which may be at the point of maximum inland penetration). Tsunami size is not simply related to earthquake magnitude. It depends on a number of factors including the depth of earthquake rupture (shallow rupture close to the trench makes for larger tsunamis) and the speed of earthquake rupture (slower rupturing events tend to make bigger tsunamis for a given earthquake size). This was first pointed out by Dr. Hiroo Kanamori, a Japanese seismologist who is currently Professor Emeritus at Caltech in Pasadena, California, in a paper on this topic published in 1972. So the 1896 event may have been a smaller, slower-rupturing earthquake compared to 2011, but its associated tsunami was similar. Since it was the tsunami that doomed the power plant, by swamping backup power and control facilities, basic common sense and sound engineering practices should have dictated that the Fukushima plant be designed to survive an event similar to the well-known 1896 tsunami.

FIGURE 4.2 Tsunami stones – historical evidence for large tsunamis in northeastern Japan. The run-up heights of tsunamis of 1896 (Sanriku earthquake) and perhaps the 869 (Jogan earthquake) may be marked by these stones. The stone engraving says "Do not build your homes below this point." Photograph by Ko Sasaki, reprinted courtesy of the *New York Times*/Redux.

There was also scientific evidence of a large earthquake and tsunami in the region in 869 AD. Dr. Koji Minoura, a professor at Tohoku University in Japan, and others published a paper in 2001 describing tsunami deposits associated with the great Jogan tsunami. He and his team carefully documented deposits associated with this event that reached 4.5 km inland, a massive tsunami very similar to the 2011 event. This important paper, published a decade prior to the Fukushima disaster, also should have given TEPCO managers sufficient evidence, and sufficient time, to beef up their backup power and cooling systems.

Clear historical evidence of very large tsunamis also exists in northern Japan, independent of scientific literature. The region is dotted with dozens of tsunami stones (Figure 4.2), marking the location of maximum water run-up in past tsunamis (Fackler, 2011c). Many of them were erected about a century ago, presumably recording

Table 4.2 *Large historical tsunamis in or near the Tohoku region*

| Year | Maximum Run-Up Height (m) | Reference |
|---|---|---|
| 869 | ? (similar to 2011?) | 1 |
| 1611 | 25 | 2,3 |
| 1896 | 38 | 2,3 |
| 1933 | 27 | 2,3 |
| 2011 | 40 | 3 |

References:
1. Minoura et al. (2001)
2. Watanabe (1998)
3. Nöggerath et al. (2011)

the run-up of the large tsunami associated with the 1896 earthquake (Table 4.2)

The combination of historical and geological evidence led to a compelling picture of rather frequent large tsunamis in the Tohoku region (Table 4.2). Unfortunately, not everyone accepted the abundant scientific and historical evidence of past large tsunamis and the implications for past large earthquakes. Earthquake hazard maps published by the government continued to show relatively low hazard in much of northern Japan as late as 2010. This raises the interesting question about what plant operators should have done in the presence of apparent scientific disagreement. Prudence would seem to dictate erring on the side of caution and accepting the safer design standard based on known tsunami hazard, rather than estimates of future earthquake size based on a discredited model employed by some government agencies. Prior to 2011, there was good evidence for tsunamis as high as 38 m in the region. Fukushima's power plant was only designed to withstand a tsunami height of 5.7 m. Moreover, unlike magnitude 9 earthquakes, these large tsunami events were not especially rare. Not counting the 2011 event, the average repeat time was less than 300 years (four events in 1064 years; see Table 4.2).

The scientific arguments indicating the likelihood of large tsunamis at the Fukushima site were brought to the attention of the TEPCO board and the regulatory agency, the Economy and Trade Ministry, at least two years before the disaster. Bloomberg, a financial reporting group, reported the minutes of a 2009 meeting where the latest findings were presented and the issue of tsunami safety was discussed. Nöggerath et al. (2011) summarize the important details of that meeting. In June 2009, Dr. Yukinobu Okamura, Director of the Active Fault Research Center of the Geological Survey of Japan, summarized risks to the Fukushima plant from large tsunamis based on the available geological data. TEPCO ignored his warnings, and government regulators refused to revise their safety standards, which would have required the company to act. Thus, the Fukushima plant met the letter of the law, even though it was clearly unsafe to many scientists and would have appeared so to even a casual observer with access to the same information. This raises troubling questions about the behavior and role of regulators, the fiduciary duty of companies to go beyond the letter of the law and follow common sense, and the widely accepted practice of excluding the public and the media from discussions related to matters of safety that affect the broader community.

Dr. Robert Geller, a professor of seismology at the University of Tokyo, has said "Utility companies were a bit cavalier when plants such as those in Fukushima were built. But as more and more knowledge came along about how dangerous they were, and they didn't upgrade the defenses, they went from being cavalier to highly negligent and/or irresponsible."

## A DISTURBING PRECURSOR

Another troubling incident, and another wake-up call for the Japanese power industry, which was unfortunately ignored, occurred in 2007. Kashiwazaki Kariwa in northwest Japan was the largest operating nuclear power plant on Earth at the time, producing 8000 MW of electricity. On July 17, 2007, it was struck by a M 6.8 earthquake, a moderate-sized event. Despite its modest size, the earthquake damaged the plant enough to force a shut-down, and a small amount of radiation was released into the Sea of Japan. The plant was offline for nearly two years, at a cost of roughly $5 B (US). Connor (2011) gives additional details.

What is surprising about this event was the fact that the plant suffered any damage at all. Earthquakes of this size or larger are common throughout Japan, and northwest Japan is frequently struck by earthquakes up to M 7.7. Yeats (2012, Figure 9.11) shows five events with M 7.3 or higher since 1940. Plants located there should be designed accordingly.

The idea that ~M 6.7–7.7 events are common in plate boundary zones around the world, and certainly in this part of Japan, was known for a long time. It is odd that a lower design standard was used at Kashiwazaki Kariwa. That standard included the assumption that the site would experience only relatively low levels of ground acceleration. The acceleration actually experienced in the 2007 earthquake was more than twice as strong as the design standard, but within the range of expected accelerations for an earthquake of this size.

Rather than assuming that the site could experience earthquakes up to M 7.7, the designers of the Kashiwazaki Kariwa nuclear plant took a different approach, known as the deterministic fault model. In this approach, active faults are first identified by detailed mapping, and their hazards are quantified based on fault characteristics such as length or total offset (long faults tend to have bigger earthquakes). The weakness in this approach is that if a fault is missed during the initial mapping phase of the study, its contribution to earthquake hazard is ignored. The geological investigations backstopping the design of the Kashiwazaki Kariwa plant were focused on land. Unfortunately the designers forgot about the offshore area, even though the plant is located near the coast. They ignored existing data that showed faulting and significant seismic risk just off the coast. This proved to be a bad decision; the July 17, 2007 earthquake occurred on an offshore fault.

Evidence for seismic hazard off Japan's west coast included a major earthquake that had struck the region just two decades earlier. In 1983, a M 7.7 earthquake occurred offshore the Tsugaru Peninsula, in the same general area as the power plant, causing a significant tsunami. Drs. Koji Minoura and Shiyu Nakaya, writing in the *Journal of Geology* in 1991, described new techniques they developed for studying tsunami deposits, techniques that Minoura would put to good use a decade later for his study of the 869 AD Jogan tsunami near Fukushima. Minoura and Nakaya used observations from the well-documented 1983 event and then compared them to earlier deposits to figure out which of the older deposits were tsunami-related. This is a classic geologic technique, summarized in the phrase "The present is the key to the past," which is usually attributed to Charles Lyell (1797–1875), a Scottish scientist and one of the founders of modern geology. Along with the principle of superposition (younger sediments over older sediments; Chapter 1), it remains one of the keystones of modern geological science.

Minoura and Nakaya used boreholes to sample the older deposits, determining that the coastal region of Tsugaru was subject to similar events every 250–400 years. Although this region is somewhat north of the Kashiwazaki Kariwa plant, the tectonic setting is similar. Prudence would dictate that the plant be designed to withstand the same higher risk. I find it surprising that Minoura and Nakaya's findings did not provoke a re-examination of seismic safety issues and a retrofit of the nuclear plant to meet the higher standard.

Given this history, you might think that officials involved with the Kashiwazaki Kariwa plant would take a more proactive attitude toward earthquake safety after 2007. If so, you would be wrong. In 2010, three years after the earthquake, with the plant back in operation, local officials decided to conduct a stress test of the newly renovated plant, assuming a joint earthquake and nuclear disaster.

However, according to a report by Funabashi and Kitazawa in the *Bulletin of the Atomic Scientists,* NISA (the Nuclear and Industrial Safety Agency) decided that such a test would unnecessarily alarm the public. Consequently, the test was performed assuming problems from snow.

## REGULATION AND TRANSPARENCY

An unfortunate aspect of our modern specialized and highly technical society is that there are a limited number of experts in any given field. This can promote a "revolving door" between the regulators (the government) and the regulated (the private sector). For experts in a given field, a successful career in government can almost guarantee a second, more lucrative career in private industry, with obvious conflicts of interest. One of the issues raised in the Fukushima and Kashiwazaki Kariwa disasters is the coziness that may have developed between the regulators and the regulated, perhaps related to this revolving-door phenomenon, limiting effective oversight. Similar concerns probably apply to many industries in many countries.

Another problem at Fukushima, and the nuclear industry of most countries, is a lack of transparency in the regulatory and enforcement process. This is less critical if government regulators are objective and have the time, resources, and motivation to perform adequate oversight. It is more important if there is possibility of conflicts of interest or lack of oversight on the part of government regulators. More transparency in Japan's nuclear industry would have allowed independent experts from academia, NGOs, and the media to evaluate the regulation, design, construction, management, and broader safety issues at Fukushima and other nuclear plants. It is usually very difficult for outsiders to obtain relevant information and perform independent oversight. Information is usually denied on grounds of business-related proprietary concerns or national security. It is difficult to see how the location of a backup water pump would give away critical industrial secrets or compromise national security, but this is usually the excuse given when the media and independent experts are denied such information. Since the safety of nuclear installations has impacts far beyond shareholder profit, business-related proprietary concerns should not be allowed to trump broader safety concerns and information access by groups conducting independent oversight. It should be possible to have independent oversight without compromising business proprietary information or security. I sometimes consult for companies and am required to sign non-disclosure agreements (NDAs) aimed at limiting disclosure of industrial secrets. Similar agreements could be crafted to allow outside review of nuclear installations.

I'll end this chapter with two questions: Should a seismic region like Japan, with its history of frequent earthquakes and tsunamis, invest so heavily in nuclear power? Would it make more sense to diversify, for example exploiting Japan's abundant active volcanoes for geothermal power?

In the next chapter, we'll look at some places where future earthquakes or tsunamis are likely, and see how well the US, Switzerland, and Turkey have applied the lessons of Fukushima. The main lesson is that not knowing or choosing to ignore the latest science on natural hazards is not only bad business, but it is dangerous for society at large. Geologists have to get the science right and communicate their findings to stakeholders. And stakeholders have to listen. Failure in either area will be increasingly costly in our crowded and interconnected world.

# 5 Future Earthquake and Tsunami Disasters

Cassandra, a member of the Greek Chorus: *"I foresee big trouble."*
Lenny: *"You're such a Cassandra."*

From *Mighty Aphrodite,* a 1995 film by Woody Allen

One of the challenges in writing a book like this is the risk of being labeled either a Cassandra or a Captain Hindsight. Cassandra was the prophet who foretold of coming tragedy and the fall of Troy (unfortunately for the Trojans, they ignored her advice about the horse). Captain Hindsight is a satirical cartoon character in the US television animated series *South Park*. Captain Hindsight parachutes into disaster zones after a disaster, lecturing the hapless inhabitants about what they should have done to avoid their current situation. I hope to avoid both labels by spending some time discussing future disasters that are likely but still avoidable through basic infrastructure investments. They are not predictions in the sense of stating a specific future time and place – the science is not good enough for that. Rather, they might be called general forecasts: With high probability, these events or problems will occur at some point in the future. Chapter 6 discusses ongoing and future hazards associated with continued use of coal. Chapter 7 discusses a more conventional hazard, flooding, including some likely future floods associated with rising sea level that will occur unless preventive action is taken. Chapter 8 reviews the environmental problems that are likely to affect our planet if we don't change the way we generate and use energy and deal with the resulting waste products. In this chapter I'll look at likely future earthquake or tsunami disasters in Seattle (US), Geneva (Switzerland), and Istanbul (Turkey). While scientists can't predict the exact timing of these events, we know with reasonable certainty where they will occur and their approximate maximum size. We also know that if we don't prepare, the outcomes could be horrific.

The timing of these future events is both a blessing and a curse. It's a blessing because we probably have sufficient time to upgrade infrastructure over the next few decades. Since everything does not need to be done at once, the rate of investment (dollars spent per year) can be modest. Some buildings can simply be replaced at the end of their life cycle with more resilient ones, using relatively straightforward improvements to construction practices, or new construction can be planned for higher ground. There is some risk to this 'go-slow' approach (the earthquake could happen next year), but given economic limitations, it's a reasonable strategy. The timing is also a curse because, given the apparent lack of urgency and the absence of a crisis, it can be politically difficult to implement necessary changes.

There are lots of major cities that lie in harm's way due to earthquake activity. Some of the larger ones include Baghdad (Iraq), Tehran (Iran), Caracas (Venezuela), Bogota (Colombia), Lima (Peru), Jakarta (Indonesia), and Manila (Philippines). I'm going to focus on two that illustrate some key points: Seattle, in the US state of Washington, and Istanbul, Turkey. I'll also discuss the risk of a tsunami that has little or nothing to do with earthquakes in the Swiss city of Geneva.

## SLEEPLESS IN SEATTLE

Imagine for a moment that an earthquake the size of the one that hit Japan in 2011 hits coastal Oregon, Washington, and southern British Columbia in the near future. While the damage in Japan was certainly major (and was greatly exacerbated by the avoidable problems at the Fukushima nuclear plant), it was not as great as it might have been. Japan's well-designed and enforced building codes meant that most buildings and other urban infrastructure escaped significant damage from earthquake shaking. In contrast, cities such as Portland, Oregon; Seattle, Washington; and Victoria, British Columbia, lacked rigorous earthquake building codes until recently. Some existing buildings do not meet this new code and would not survive such an event. Major businesses such as Microsoft and Boeing could be greatly affected, potentially hurting the global economy. If such an earthquake occurred in the near future, casualties in the region could easily overwhelm emergency facilities. A more optimistic scenario is that this earthquake (which is almost certain to occur at some point in the future) will not strike for at least a few more decades, and it may not strike for a century or more, giving these regions and nearby towns and cities a window of opportunity to ensure that their infrastructure becomes more earthquake-resilient.

The geological background is straightforward. Seattle – as well as cities such as Portland, Oregon, to the south and Victoria, British Columbia, to the north – sit near a subduction zone (Figures 2.3 and 2.4), with an oceanic plate offshore to the west (in this case the Juan de Fuca Plate) subducting beneath the western margin of North America (Figure 5.1). As in most subduction zones, the process causes a series of volcanoes to form inland from the coast and can cause big earthquakes, which in some cases are accompanied by big tsunamis. In the US, the volcanic chain is called the Cascades Volcanic Arc, and the subduction zone is often referred to as the Cascadia subduction zone. Washington and Oregon, parts of northern California, and parts of southern British Columbia in Canada (including Vancouver Island and the provincial capital, Victoria) are part of this subduction system.

One question the informed reader might ask is, if cities such as Seattle and a Japanese city such as Tokyo are both located near well-known subduction zones, why is Tokyo's infrastructure more earthquake-resilient than Seattle's? The answer is quite interesting, and has to do with the history of European settlement, the more recent history of the discovery of plate tectonics, our associated understanding of earthquake process and risk, and the concepts of sticky infrastructure and time lag first discussed in Chapter Three.

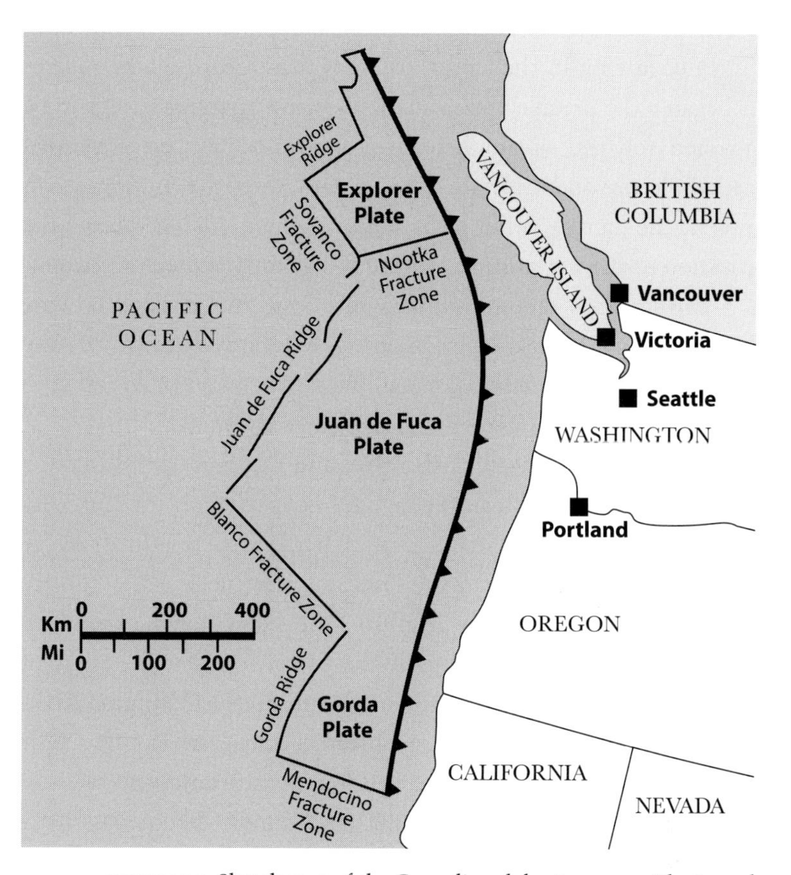


FIGURE 5.1 Sketch map of the Cascadia subduction zone. The Juan de Fuca Plate, offshore the American states of Washington and Oregon and the Canadian province of British Columbia, moves northwest at about 4 cm/yr, subducting (pushing) under the west coast of North America. The smaller Gorda Plate to the south and Explorer Plate to the north also subduct. The heavy line with teeth shows the subduction boundary between the plates and the approximate location of the sediment-filled trench.

## *Settlement History*

Japan's recorded history goes back more than 1000 years. There are detailed descriptions of earthquakes, tsunamis, volcanic eruptions, and other natural events, including location, size, and impact on the population. Ancient earthquakes can be compared to modern ones (the ones occurring in the last ~120 years, when earthquake magnitudes began to be precisely measured) allowing earthquake size to be determined for the entire historical record. This has generated a detailed picture and understanding of natural events. As we saw in Chapter 4, the picture is not perfect: Some scientists interpreted the information to suggest northeast Japan would only experience magnitude 8 earthquakes rather than magnitude 9, but it does provide the basis for a well-designed and rigorously followed set of building practices that ensure earthquake resilience. Tokyo is not immune to massive earthquakes, but it will survive and survive well.

In contrast, European civilization and the practice of making a written record of significant events did not arrive on the Cascadia margin until recently. British exploration began with the third voyage of James Cook, who explored the Cascadia margin in 1778, shortly after visiting the Hawaiian Islands. The Vancouver expedition explored a considerable portion of the northwest Pacific coast of North America and sailed more than 100 km up the Columbia River in 1792. The Lewis and Clark expedition reached the Pacific Coast near what is present-day Portland in 1805. Significant settlement of Seattle did not begin until after 1850. No major earthquakes have struck the Cascadia margin since 1700, so they are not part of the public consciousness.

### *Scientific History*

Scientific understanding of the earthquake process here is also recent. In the early days of plate tectonics, in the 1960s and 1970s, when our understanding of where, how and why earthquakes occurred was still relatively new, some subduction zones that had not experienced a major quake in recorded history were thought to be "aseismic." In other words, the interface between the plates was thought to have sufficiently low friction that strain might not accumulate and would instead be released gradually in many small movements so that major earthquakes would not occur. The Cascadia subduction zone was a possible candidate, as it had not experienced a major earthquake in recorded (Western) history. But most scientists recognized at the time that this history was woefully short, less than 200 years.

Other aspects of the geology of the region also appeared to be consistent with the lack of earthquakes. For example, most subduction zones that produce large and great earthquakes have deep trenches offshore, marking the surface location where the downgoing plate first begins to bend as it dives beneath the upper plate (Figure 2.4). But the Cascadia margin lacks such a trench. This absence, combined with the lack of earthquakes, led some scientists to propose that the Cascadia margin was no longer an active subduction zone; perhaps the two plates had recently become welded together, moving as one, eliminating the possibility of a major subduction zone earthquake.

The recognition that the Cascadia margin is in fact an active subduction zone, capable of producing large great earthquakes and large tsunamis, similar to the 2011 events in Japan, has come only in the last few decades. How this happened is a fascinating detective story.

One part of the puzzle that was easily answered was the apparent lack of a trench. Seismic reflection (an acoustic imaging technique described in Box 5.1 and Figure 5.2) showed that there actually is a trench; it's just not obvious in simple bathymetric charts. The reason is that the trench is filled with sediment. What this means in geological terms is that the rate of sedimentation (how fast the trench fills with sediment) is faster than the rate at which subduction creates the trench by dragging down the upper plate. High rainfall in the region and easily erodible volcanic mountains both play a role, dumping lots of sediment into the Fraser, Skagit, and Columbia rivers. The rate of sedimentation was also higher in the recent past. For example, small glaciers currently cover some of the higher peaks of the Cascades and contribute to the erosion of these mountains. But starting about 100,000 years ago and up until about 8,000 years ago, these glaciers were much more extensive and were extremely effective

### BOX 5.1 **Seismic reflection and medical X-rays**

Seismic reflection is similar in some ways to medical X-rays. Both techniques use backscattered energy to produce images. While X-rays use high-energy electromagnetic radiation, seismic reflection uses acoustic energy, low-frequency sound waves produced by small explosions, or air guns. Figure 5.2 shows a beautiful example from Lake Geneva in Switzerland. The sound energy bounces off different layers of sediment, and energy from deeper layers arrives later. The vertical scale in Figure 5.2 represents this and is labeled as two-way travel time (TWT, i.e. the time it takes for the wave to travel to the layer and to be reflected back to the surface) in milliseconds (thousandths of a second). Since we know the approximate speed of sound in water and sediment, the time differences can be translated into depth differences, allowing a virtual cross section to be generated. However, the conversion factor from time to depth is not precisely known, so it is traditional to keep the vertical scale in terms of time (the thing that is measured) rather than distance (the thing that is estimated). The sediment section in Figure 5.2 is about 30–40 meters thick and lies about 300 meters below the lake surface.

at carving out vast amounts of sediment and moving it downhill to the coast and beyond.

In 1981, Dr. Jim Savage, a scientist at the US Geological Survey – with colleagues Dr. Michael Lisowski and Dr. William Prescott – published a paper analyzing a decade's worth of trilateration data from coastal Washington. Trilateration measures the distance between markers set firmly into the ground. The data showed that the distance between markers oriented at right angles to the coast had decreased in the previous decade, a clear indication that the plate boundary was locked, causing the upper plate to deform (squeeze together) and accumulate strain to be released in a future earthquake. Since it's likely that strain had also been accumulating prior to the

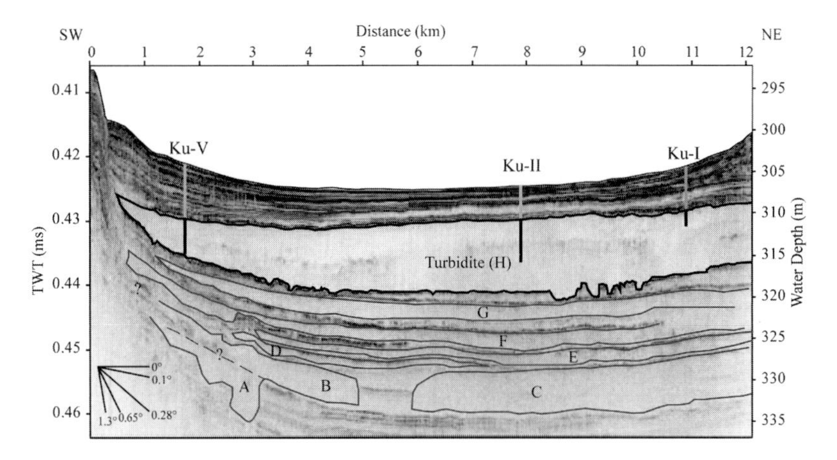


FIGURE 5.2 A seismic cross section (Box 5.1) from Lake Geneva in Switzerland, from work by Katrina Kremer and colleagues at the University of Geneva. The cross section covers a distance of 12 km (about 7.5) miles). The vertical scale is depth, estimated from the two-way travel time (TWT) of the acoustic signal in milliseconds (1 ms = 1 thousandth of a second). Water depth here is about 300 meters (1,100 feet). The lens-shaped unit labeled H is a deposit known as a turbidite (see Box 5.2) caused by a large landslide and tsunami in 563 AD that destroyed a town at the western end of the lake where the modern city of Geneva is now located. The tsunami was probably caused by an underwater landslide from the Rhone River delta at the eastern end of lake, when the delta slumped abruptly into deeper parts of the lake, forming the turbidite (see next section, Tsunamis in Switzerland). Vertical lines mark the location of drill cores used to recover sediment samples for later study and dating. Similar units below the youngest turbidite deposit are labeled A through G and tend to be thinner, in part because of sediment compaction. These older units suggest that tsunamis have occurred repeatedly in the past. Figure courtesy of K. Kremer (personal communication). See Kremer et al. (2012, 2014).

measurements (and probably at about the same rate) and since no earthquakes had occurred in recorded history to release this strain, the simplest interpretation was that the region would experience a large earthquake at some point in the future. In the conclusion to their paper, the scientists wrote, "The implication is clearly that the Washington and Vancouver Island coasts are subject to great, shallow, thrust earthquakes."

In a perfect world, the significance of this report would have been immediately recognized. Cities near the coast would immediately enact strong building codes, much as Los Angeles did after the 1933 Long Beach earthquake, and today, more than 30 years later, much of the urban infrastructure in the region would be much safer. As with most "first warnings," however, Savage and his colleagues' report was largely ignored outside a few scientific circles. This reflected, in part, the fact that most people were not familiar with trilateration data or the science of geodesy. Even many scientists were unfamiliar with trilateration and the rather complex calculations required to interpret the data. Another factor was the apparent lack of urgency. Savage and colleagues were not reporting an earthquake or predicting one in the next few years, just warning of one in the possibly distant future.

The next piece of the puzzle came more than five years later. Dr. Brian Atwater, a geologist with the US Geological Survey (and an author on one of the tsunami papers in Japan cited in Chapter 4), had been studying coastal deposits in westernmost Washington. In 1987, he published a paper describing coastal marsh deposits that had been abruptly buried and covered with a sheet of sand. Part of his study involved trenching and the use of the superposition principle described in Chapters 1 and 4 (younger sediments normally lie above older sediments). He interpreted the burial as subsidence due to an earthquake (see Box 2.4, middle panel), and the sand sheet as the subsequent tsunami deposit. He noted that this sequence had occurred several times in the last few thousand years, about the frequency expected if these events were caused by great subduction earthquakes.

But when exactly was the last earthquake? Dr. David Yamaguchi, then at the University of Colorado, and his colleagues looked at evidence from tree rings and radiocarbon dating of drowned trees. In papers published in 1991 (Atwater and Yamaguchi, 1991) and 1997 (Yamaguchi et al., 1997), they showed that the last such event was a little more than 300 years ago, near the end of the 1600s.

The final piece of the puzzle came, paradoxically, from Japan. Dr. Kenji Satake and colleagues discovered in Japan's precise earthquake and tsunami records evidence for a massive tsunami on January 26, 1700, but there was no corresponding record of any earthquake. Nor were there any records of major earthquakes in the other possible locations that would impact Japan (Kamchatka, Alaska, or South America). The best candidate would be Cascadia, but only if it was a magnitude 9 event, rupturing up to 1000 km of the Cascadia margin. By this time, the coastal deposits and drowned trees first reported by Atwater had been well mapped, and it was known that the latest event had in fact ruptured that length of margin. Tree-ring dating had also improved in precision, and it was known by then that the latest Cascadia mega-earthquake had to have occurred between 1699 and 1700 AD (Jacoby et al., 1997).

How soon might the next great Cascadia event occur? To answer that question, we would need a precise record of many past events, not just the last few, to get an accurate idea of recurrence interval and pattern. Do mega-earthquakes occur regularly, e.g. every 500 years, or do they cluster in time, with several earthquakes in quick succession followed by long periods of quiescence? The tree-ring record can only take us back so far (few trees live beyond 500–1000 years in the humid coastal environment of the US Pacific Northwest), and marsh deposits can also be difficult to interpret as we go farther back in time. We can gain additional perspective by looking offshore at a special type of underwater landslide deposit often associated with great earthquakes. The sediments resulting from these catastrophic slope failures are very distinctive and are called turbidites. They had been studied by geologists for many years before their origin was fully understood (Box 5.2).

Turbidite flows can happen randomly whenever a river delta or other locus of sedimentation becomes too steep or when a pulse of new sediment from a major storm or shaking from an earthquake causes an already steep slope to fail (see the next section on Lake

### BOX 5.2 **Turbidites and Bouma sequences**

In 1962, a geologist named Arnold Bouma published a classic book on the origin of turbidites. These layered sedimentary deposits are found all over the world in rocks of many different ages. A typical turbidite has a layer of sand, followed by silt (finer-grained sand), and finally mud. The sequence is so common in the geologic record that it has become known as a Bouma sequence. Bouma realized that the sequence was a characteristic deposit from a turbidity current, a turbulent, gravity-driven flow composed of a slurry of mud, silt, and sand. It starts out as an underwater landslide or debris flow on a steep slope moving chaotically downhill. As the slope decreases and eventually flattens out, the larger blocks slow down and stop, but a slurry that includes finer particles, including sand, silt, and mud, can continue for many kilometers. Eventually this also slows down, and as the flow finally comes to rest, the nature of fluid flow changes from "turbulent" to "laminar." Scientists who study fluid flow like to distinguish between these two types of flows: Laminar refers to smooth flow that is typical of slower-moving fluids over flat terrain or in a smooth-sided pipe, while turbulent refers to messy, unorganized flow that is typical of fast-moving fluids over irregular terrain or a rough-sided pipe. Think of a canoe on a slow-moving river versus a kayak on white water.

As the turbidite flow slows and becomes laminar, heavier sand particles fall out of suspension first. Later, finer-grained silts and muds settle out: first the silt and then finally the mud on top. Geologists call this graded bedding. It's easy to recognize in an outcrop (a large area of rock that is exposed, or "crops out," at the Earth's surface and is beloved by geologists for what it tells us about the Earth). Graded bedding is one of the classic signs of a Bouma sequence. Generations of undergraduate geology students have attended field camps, stood on outcrops to inspect graded bedding, and been tutored on the connections between Bouma sequences, the laws of fluid dynamics, and the mysteries of laminar versus turbulent flow.

Geneva). It is this last possibility that can be exploited to investigate earthquake recurrence intervals.

The connection between turbidites and earthquakes was first recognized after the 1929 Grand Banks earthquake in Newfoundland. This event caused significant local damage, mainly from a subsequent tsunami. The occurrence of the tsunami was a little strange because it was not in a subduction zone. But in the hours after the earthquake, something even stranger happened. A submarine telegraph cable offshore Newfoundland was severed, miles away from the earthquake. Some time later, communication on a second cable, farther away and in deeper water than the first, was also lost. It took several decades before the sequence of events was fully explained. In 1952, two scientists at Lamont-Doherty Geological Observatory (LDGO) in New York, Drs. Bruce Heezen and Maurice Ewing, realized that an underwater landslide caused by the earthquake had developed into a large, rapidly moving turbidity current, severing the cables sequentially as the turbidite moved downslope. The timing of the cable breaks had been recorded, allowing the direction and speed of the turbidite flow to be determined. Later studies in other parts of the world confirmed the link between turbidites and earthquakes (e.g. Houtz and Wellman, 1962; Sims, 1975).

It's obviously not possible to dig a trench to study young turbidite deposits, since the critical layers are at the bottom of the ocean or deep lakes. These underwater sediments can instead be sampled from a ship by using specialized sampling techniques from the field of marine geology such as box coring (which samples roughly the upper meter, but leaves the surface intact), piston coring (which can sample up to 10 meters of sediment core, but may disturb the surface), or seismic reflection (which produces a cross section image tens to hundreds of meters or more beneath the sea floor) (Box 5.1 and Figure 5.2).

It took several decades after the Heezen and Ewing study before the connection between turbidites and earthquakes was fully exploited to study earthquake recurrence intervals. In 1984, another scientist at LDGO, Dr. Kim Kastens, published a paper based on her earlier Ph.D. research at Scripps Institution of Oceanography in California. She proposed that a particularly widespread series of turbidite deposits in the eastern Mediterranean with similar ages was best explained by a large earthquake trigger (otherwise the timing of the deposits would tend to be random). She used the timing of the deposits to estimate a recurrence interval (1,500 years) for the causative earthquakes. About the time Kastens was publishing her work, Dr. John Adams, a seismologist with the Geological Survey of Canada, was studying turbidites offshore Oregon and Washington using cores collected earlier by Garry Griggs, a student working with Dr. LaVerne Kulm at Oregon State University (OSU) that had been stored at OSU's core lab. Adams published his work in 1990. These and other studies done around the same time effectively established the field of turbidite paleoseismology: using turbidites to estimate earthquake recurrence intervals.

Adams' 1990 paper was the first to link Cascadia turbidites with great earthquakes in the region. He showed that a detailed record existed in sediment cores and demonstrated that earthquake-related tsunami deposits occurred along nearly 600 km of the Oregon-Washington margin (meaning that the earthquakes must have been very large to affect such a large area). He found 13 events younger than 7,700 years, defining the average time between great earthquakes as approximately 590 years (7,700 divided by 13). The age assignment is based on the presence of distinctive glass shards from a volcanic eruption known to be about 7,700 years old. These shards mark the eruption of Mt. Mazama, a volcano that was located at the present position of Crater Lake in Oregon. The eruption was so large that it blew the top off the mountain, creating the present lake and distributing volcanic material over a large fraction of western North America, including the offshore Cascadia margin. If you've ever seen Crater Lake, you'll know why geologists cringe whenever we hear politicians talk about cutting the funding for volcano monitoring. To geologists, 7,700 years ago is like yesterday. If it happened yesterday, there's a reasonable chance it could happen tomorrow.

Can we use 590 years to deduce the timing of the next earthquake? In other words, are we safe until 2290 AD (1700 AD + 590 years)? No, for at least two reasons. First, there is some uncertainty in the age estimates for the turbidites. Second, it depends on how periodic the earthquakes actually are. In Adams' 1990 paper, the thickness of clay layers between the earthquake turbidites was described as quite variable, implying the possibility of varying time spans between earthquakes. In 1997, Atwater and Hemphil-Haley described a 3,500 year record from Willapa Bay, Washington that also suggested variable timing between earthquakes. In 2003, Dr. Chris Goldfinger, a marine geologist at OSU, and colleagues extended Adams' work, publishing a remarkable 10,000 year turbidite record of major earthquakes off the Cascadia coast that greatly clarified the question of earthquake periodicity.

Goldfinger started this work in 1996 with colleague Hans Nelson. They were initially skeptical of Adams' work, assuming that past earthquakes would be difficult to discern from the turbidite record here due to the complexity of the channel systems, storm deposits, and other factors. But after spending a month at sea and collecting new cores, they realized that Adams was right. In an email, Goldfinger told me, "It's a neat story, and frankly taught me a lot about how science should work, and how not to hold into ideas too tightly, but tightly enough to give them a fair shot. That John worked this out is an enduring example of elegant science." In another email comment, Adams told me, "It's a wonderful real-world example of how science should work."

Goldfinger's work documented 18 major events, 13 of them above (younger) than the Mt. Mazama ash. The sediment cores studied by Goldfinger and colleagues also show a fair amount of variability in the time between earthquakes. In other words, the earthquakes are not strictly periodic, and we can't count on another few hundred years of safety. Updated analyses (Goldfinger et al., 2012, 2013) suggest highly variable intervals ~200–1200 years in length, with evidence for earthquake clustering. The largest earthquakes tend to occur at the end of a cluster. Given these findings, the prudent thing to do is to use the next few decades to refine the turbidite record and other relevant data, and simultaneously increase earthquake resilience in the urban areas of the Cascadia margin. This is happening to some extent, but there are still a large number of unreinforced masonry buildings, especially in the older urban cores of big cities like Seattle and Portland. Experience in other cities hit with major earthquakes shows that these buildings are especially susceptible to catastrophic collapse, even for earthquakes of moderate size. Hopefully these can be upgraded or replaced soon.

Thompson (2012) has written an excellent book on the Cascadia earthquake and tsunami for readers interested in this fascinating geological detective story.

## TSUNAMIS IN SWITZERLAND

In the previous section, we looked at a case where turbidite deposits were used as an indicator of past earthquakes. Earthquake shaking stimulated an underwater landslide, and the resulting turbidite deposit helped to determine the timing of the event, using the principle of superposition and various dating techniques. The associated tsunami was also caused by the earthquake. The turbidite deposit and tsunami occurred at about the same time but were not directly related; both were consequences of the earthquake.

But turbidites and tsunamis can be caused by things besides earthquakes. Whenever a large underwater landmass moves quickly, a tsunami can be generated. In this section, I'll describe an example from Lake Geneva in Switzerland, where an underwater landslide at the east end of the lake, probably unrelated to an earthquake, caused a tsunami. The resulting turbidite deposit in the bottom of the lake near its east end can be used to study future tsunami risk to Geneva.

Historical records suggest that a tsunami devastated a town at the southwest end of Lake Geneva, the current location of the city of Geneva, in 563 AD. Dr. Katrina Kremer and colleagues at the University of Geneva used seismic reflection and sediment core data from the lake bottom, combined with some common sense inferences about the regional geography, to figure out the source of the tsunami. In this case, the source is probably an underwater landslide from the Rhone River delta at the east end of the lake. Deltas are discussed in more detail in Chapter 7, but basically, they are piles of sediment (sand, silt, and mud) that form at the mouth of a river. As the delta accumulates sediments from the river, its slope becomes too steep to remain stable, and the sediments cascade down the face of the delta into deeper parts of the lake, forming turbidite deposits similar to those produced by earthquakes in other parts of the world. Sediment cores taken at the same location as the seismic reflection imagery shown in Figure 5.2 allow the sequence of sediments to be dated (remember superposition), bracketing the age of the tsunami deposit and allowing the geological record to be correlated with the historical record. Modeling by the Geneva group suggests that the underwater landslide reached a speed of about 50 km/hr (~30 miles/hr).

The astute reader may notice that in Figure 5.2, the sediment cores only penetrate the upper (youngest) turbidite deposit, but there are at least seven deeper (older) turbidite deposits. In other words, these events have happened many times in the past. Unfortunately, information on the timing of these earlier events, an important factor in risk assessment, is not yet available. Drilling into the deeper units and dating them with radio-carbon or other techniques will give clues to the frequency of such events and could help determine whether Geneva is at risk of another tsunami in the near future. Such studies are currently planned (Kremer et al., 2012).

The astute reader may also notice that given the early warning from this study, there is time for mitigating measures. Zoning changes in Geneva could promote infrastructure change in low-lying areas near the lakefront, such as conversion to green space. Sensors on the delta or lake bottom could also be deployed to give city residents several minutes warning of an imminent tsunami, allowing evacuation to higher ground.

### THE ORIENT EXPRESS

In 1883, railroad service between Paris and Constantinople (the old name for Istanbul) was initiated and became known as the Orient Express. The service would eventually become famous, romanticized in novels such as Agatha Christie's *Murder on the Orient Express*. Initial runs were arduous for travelers, especially before 1889, involving several changes of train, transfers to and from ferries, and, depending on equipment reliability, several weeks of travel. Progress toward Istanbul could be painfully slow. Train service uninterrupted by equipment changes began in 1889. However, even when everything was working, delays due to weather would occur. One train was delayed in Turkey for six days due to snow (a key plot point in Christie's novel). On longer time scales, the service was also irregular. The route changed several times, and service could be interrupted for long periods, for example by World Wars I and II, or more recently by economics, as air travel provided stiff competition for long distance train travel, forcing cancellation of the service for several years.

On December 26, 1939, a large (magnitude ~7.9) earthquake struck eastern Turkey, killing tens of thousands of people, and leaving many people homeless during cold winter conditions. World War II had started in Europe, and international relief efforts were slow to non-existent. The earthquake, which became known as the great Erzincan earthquake, occurred on the North Anatolian fault. This fault is similar in many respects to the San Andreas Fault in California and separates the small Anatolian Plate to the south from the much larger Eurasian Plate to the north. The plates move sideways past each other at a rate of about 2 cm/yr, with the Anatolian Plate moving west relative to Eurasia (Figure 5.4). Over the next five years, between 1942 and 1944, three additional major earthquakes occurred on the North Anatolian Fault, each one west of the previous one, getting closer and closer to Istanbul.

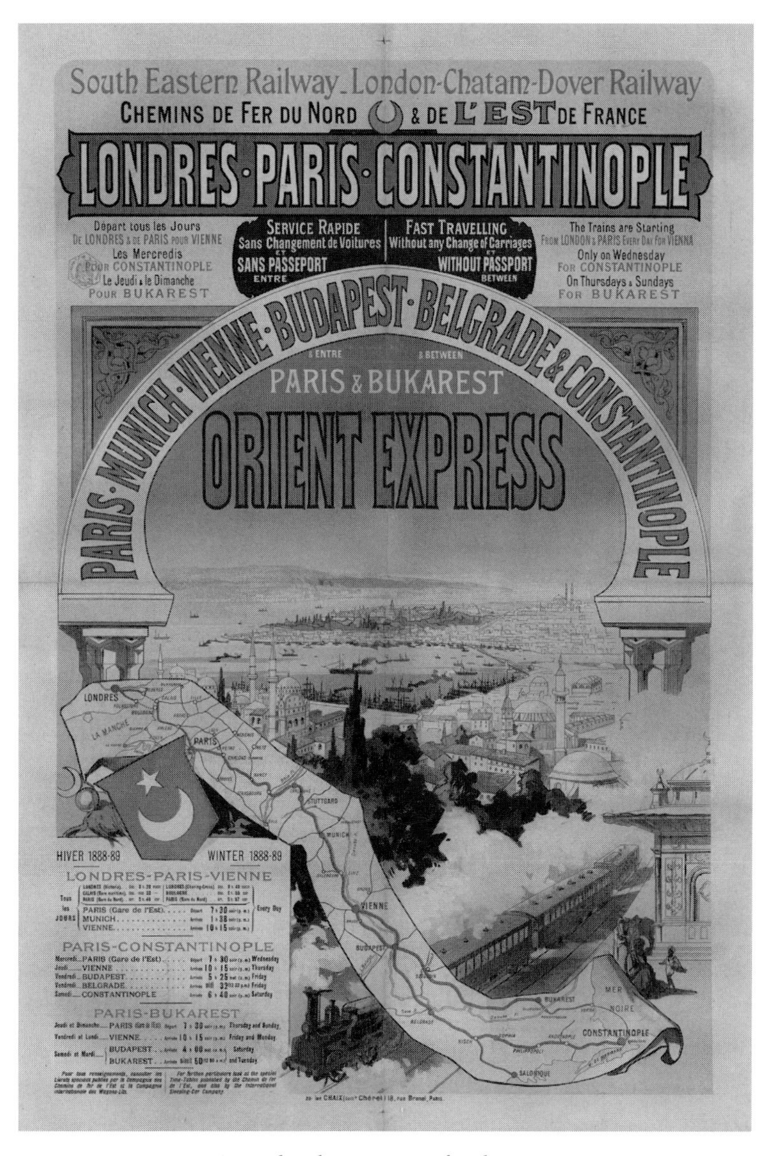


FIGURE 5.3 An early advertisement for the Orient Express. An 1888 advertisement for the Orient Express, from a poster by French lithographer Jules Cheret. Figure courtesy of Wikipedia Commons.

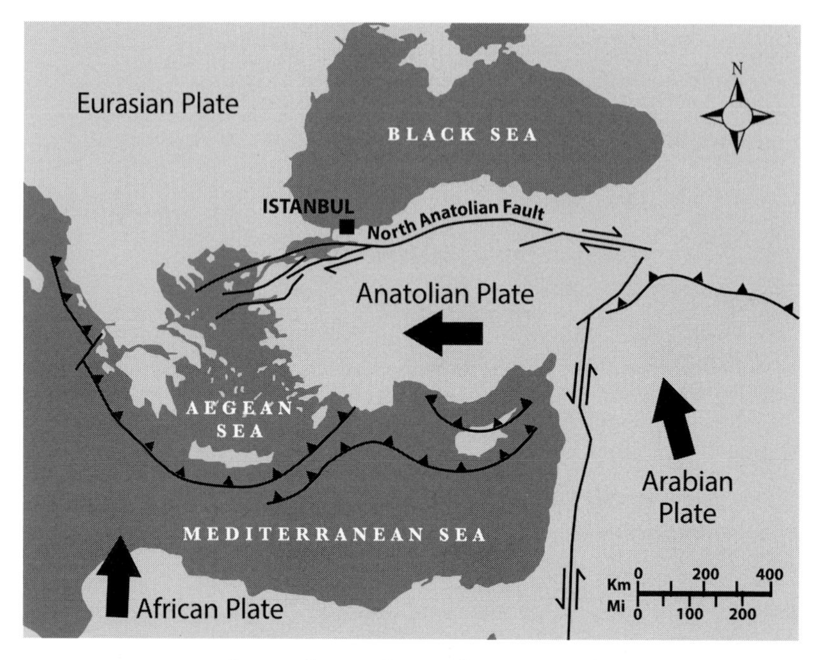


FIGURE 5.4 The North Anatolian Fault represents the northern boundary between the Anatolian Plate and the Eurasian Plate. The thick arrows show the approximate direction of movement of the various plates in the region (African, Arabian, and Anatolian) relative to the Eurasian Plate. Note that Istanbul lies close to the North Anatolian Fault. McClusky et al. (2000) and Uzel et al. (2010) give additional details.

This westward migration of earthquakes in a short period of time is perhaps the most famous example of a propagating earthquake sequence and has been studied by many scientists. With major earthquakes occurring nearly every year for five years (1939, 1942, 1943, and 1944) and with each one west of the previous one, it seemed at first like earthquakes in the region could be forecast with a high degree of assurance. Ihsan Ketin, a Turkish geologist, published the first modern study of the sequence in 1948. He mapped the North Anatolian Fault, and clearly showed the city of Istanbul in harm's way near its western end. Charles Richter, inventor of the Richter scale for earthquake magnitude, discussed the sequence in his classic textbook, *Earthquake Seismology*, which was first published in 1958.

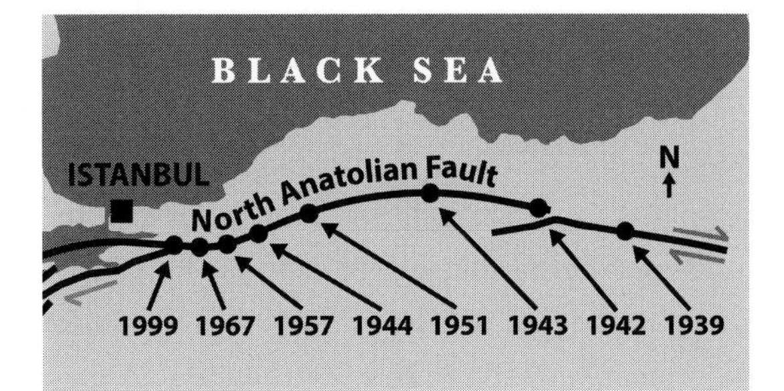


FIGURE 5.5 This map of the North Anatolian Fault shows rupture locations of large earthquakes on this fault during the last century. Note the location of Istanbul. Large earthquakes are defined as those having more than 2 meters of slip, using the data of Pondard et al. (2007).

Scientific interest in the earthquake propagation process waned to some extent over the next few decades as the simple progression stopped. The next event did not occur until 1951. It was a much smaller event and was not located west of the previous earthquake. Instead, it was located between the 1943 and 1944 events (Figure 5.5). Like the Orient Express, the progression of earthquakes toward Istanbul was somewhat irregular. This has caused problems in terms of focusing attention on earthquake hazard in the city. The next two events, in 1957 and 1967, continued the slow westward progression toward Istanbul, but they were also relatively small events.

In 1997, Dr. Ross Stein, a seismologist at the US Geological Survey, with colleagues Aykut Barka and James Dieterich, published a landmark paper that went a long way toward clarifying what was happening on the North Anatolian Fault. Stein and colleagues assumed that the probability of a future earthquake is related to the long-term stress increase on a fault due to plate motion combined with a short-term stress perturbation due to recent earthquakes nearby. Depending on the fault orientation and the type of earthquake, the short-term stress perturbation can either increase or decrease the amount of stress that promotes an earthquake, increasing or decreasing the corresponding probability of the next earthquake in a given time frame. By taking the known amount of slip that occurred in each of the ten earthquakes that had ruptured various segments of the North Anatolian Fault between 1939 and 1967, the known orientation of those fault segments, and the background "loading rate" from plate motion (2 cm/yr) determined by high precision GPS, the corresponding stress changes on adjacent fault segments can be calculated. This stress gain or loss can be translated into a probability increase or decrease for an earthquake. Stein and colleagues found that in nine out of ten cases, the last earthquake raised the probability for the next earthquake on the adjacent segment. While this explanation and forecasting technique might not satisfy everyone (for example, it could not lead to precise predictions of when and where the next earthquake would occur), it should be extremely useful for engineers and planners, as it gives information on where and when to focus scarce resources for earthquake preparedness.

Stein and colleagues also issued an ominous warning. In the second to last sentence of their summary, they stated: "During the next 30 years, we estimate ... a 12 per cent probability for a large event south of the major western port city of Izmit."

In fact, the major earthquake at Izmit that Stein and colleagues forecast occurred two years later in 1999, a strong confirmation of the technique well within their prediction window. At least 17,000 people died in the Izmit earthquake, and more than 300,000 people were left homeless, a significant fraction of the region's population. As with many previous warnings of impending disaster in the scientific literature, it appears that civil authorities in Izmit did not pay too much attention to Stein and his colleagues' publication and forecast. On the positive side, unlike the 1939 event, more than a dozen countries sent aid.

One avoidable aspect of the damage concerned a series of petroleum storage tanks with limited fire suppression capability that were situated close together and close to the fault. One caught fire shortly after the earthquake, and authorities were unable to put out the flames. Over the next few days, a total of 17 adjacent storage tanks caught fire, causing considerable damage. Military authorities learned long ago to store explosives in a series of small bunkers spaced far apart so that an explosion in one would not set off a larger disaster. Industrial facilities with flammable or explosive materials should follow the same logic.

What about future danger to Istanbul, a much bigger city than Izmit, with a population of approximately ten million people? Several studies suggest a M > 7 event is likely in the near future. Dr. Mustafa Erdik, Professor of Earthquake Engineering at Bogazici University in Istanbul, puts the probability at 2 percent per year and has said, "İstanbul has the highest probability of an earthquake among the world's mega cities."

Stein and colleagues compared the initial four 1939–1944 earthquakes to falling dominoes. I find the comparison to the longer (1939–1999) sequence also appropriate, perhaps more so than the authors intended. The few times I have tried to set up dominoes, I didn't have much luck at getting them to fall in a nice sequence. Sometimes there was an interruption, and I had to "intervene" to get an especially poorly behaved group to continue. Sometimes they fell in the "wrong" direction. Like my recalcitrant dominoes and the Orient Express, "on-time arrival" of an earthquake in Istanbul is not guaranteed. It is not even certain that the next earthquake on the North Anatolian Fault will be west of Izmit, or anywhere near Istanbul. But there is a reasonable probability that it will.

One way to look at Istanbul's odds is to simply consider history. Twelve major earthquakes have damaged Istanbul in the last 1,500 years, giving an average recurrence interval of about 125 years. Istanbul last experienced a significant earthquake in 1894, suggesting that the city is once again entering the time window for a major event.

Istanbul has made significant progress in hardening its infrastructure since the 1999 Izmit earthquake. Nevertheless, a 2009 study found that 30 percent of structures in the metropolitan area remain weak.

Up to now, I've focused on things that are obvious natural hazards, such as earthquakes and hurricanes, and discussed some common sense solutions to the associated infrastructure problems. The next three chapters look at environmental issues not usually discussed in the same context as natural hazards. Unlike the "bang" of an earthquake, they are slow-moving "whimpers" like ongoing pollution from coal (Chapter 6) or sea-level rise (Chapter 7). Solutions to these issues tend to be more complex and also politically controversial. The issues I'll discuss in these chapters nevertheless have similarities with natural hazards, including the role of communication (or its lack) among scientists, policy makers and the public, and the importance of taking the long-term view. I hope to convince the reader of these similarities, and also convince the reader that, as with natural hazards, many environmental issues also have common sense infrastructure solutions.

# 6 Nuclear Power, Coal, and Tuna: The Concept of Relative Risk

*"Follow the money."*

Deep Throat, in the 1976 American film *All the President's Men*

From Chapter 4, it might appear that I am opposed to nuclear power. In fact, I think nuclear has an important role to play in our mix of power sources, at least for the next few decades until we develop economic forms of clean, renewable energy. The arguments against nuclear power mainly involve short-term radiation risk from accidents and the cost and risks associated with long-term storage of waste. These problems can be minimized by designing and building smarter. However, risk can never been eliminated completely, and this makes many people quite nervous.

The risks and costs associated with nuclear power should not be evaluated in a vacuum but instead need to be compared with the risks and costs of other forms of power generation. When it comes to energy, there is no free lunch. If you've ever turned on a light bulb in a darkened room, you are part of a large energy-user community, and this places some limits on your right to complain about the environmental risks and costs of power generation. But you do have a right to ask if we are proceeding in a logical, cost-effective, and safe way, one that minimizes risk of environmental harm, including harm to future generations. This brings us to the concept of relative risk, related to the concept of cost to benefit ratio first discussed in Chapter 3. Relative risk is an important part of the discussion of natural and human-made hazards. This is especially true when we focus on the costs of mitigation and best practices. In this chapter, I'll investigate the concept of relative risk by comparing the environmental costs of nuclear power with another common energy source: coal. I'll attempt a quantitative comparison of nuclear and coal-fired power plants in terms of relative health and mortality risk, discuss some related environmental issues, and look at how the media informs the public about these issues in the face of powerful special interests.

Coal has a host of environmental problems associated with its mining, transport, combustion and associated air pollution, and subsequent groundwater pollution associated with fly ash, its main waste product. I'll discuss the problem of mercury, a dangerous neurotoxin that is emitted during coal combustion, and make analogies with lead, another neurotoxin that has been better studied. I'll also look briefly at costs associated with solar power, a possible alternative to coal and nuclear power in some locations.

Coal and nuclear fuel produce electricity in pretty much the same way: They heat a fluid (e.g. water) to make a higher volume phase (steam) to drive a turbine. They are more sophisticated than the steam-driven engines first built commercially by Thomas Newcomen around 1712, but the principle is the same. The early engines were first used to pump water out of mines, enabling deeper mining of coal and various metals, and were later improved upon by James Watt. Watt's design included a separate condenser for the steam to improve efficiency (the condenser saves some of the heat originally used to make the steam). His design was commercialized by 1775 through an alliance with businessman Mathew Boulton. The Watt-Boulton engine was extremely successful and essentially started the industrial revolution and, with it, widespread use of coal.

Making power by heating a fluid with either coal or nuclear fuel can be done safely and efficiently while minimizing environmental harm, assuming sound engineering and management. The two approaches to electricity production differ principally in the nature of their environmental impacts and in public perceptions of their relative risk. It should be noted that while both approaches to power production are currently profitable without direct subsidy, this profitability depends in part on being able to offload some of the environmental costs to other stakeholders, including future generations. To do a meaningful comparison, we need to examine those costs in more detail. They are unfortunately not considered in standard economic analysis, a problem I'll return to later.

Some of the problems with nuclear power were considered in Chapter 4. One thing not touched on in that discussion is the long-term cost of storage for spent fuel. This is a contentious topic. In the US, a repository constructed for this purpose in Nevada has never been used because of political pressure, even though it represents a very large national investment. Hence it is difficult to calculate true long-term costs; it depends in part on future political decisions about whether or not to use the existing repository. Many environmentalists are opposed to nuclear power because of the uncertainties and risks associated with long-term storage of spent fuel rods and other high-level nuclear waste. My feeling is that these risks are relatively low (certainly compared to other risks discussed below) and, in the absence of political considerations, the technical challenges are surmountable at reasonable cost. Coal also has long-term waste disposal issues. For example, it produces huge volumes of fly ash, a waste by-product of combustion that can impact freshwater resources depending on how and where it is stored. In general these long-term costs are also poorly accounted for.

Environmental issues related to large-scale mining, transport, burning, and waste disposal of coal are considerable, but they don't always receive the same attention as nuclear power. The energy density of coal is much lower than uranium, meaning that much larger volumes of coal need to be mined, transported, and burned, and much larger volumes of waste disposed of, compared to nuclear fuel. Coal mining often involves strip-mining and deforestation of huge tracts of environmentally sensitive areas (Figure 6.1). In some cases, this can impact water resources due to acid-mine drainage. In Pennsylvania, coal has been extracted in underground mines for more than a hundred years, leading to miles of abandoned tunnels that sometimes cause subsidence in overlying regions and damage to buildings. Coal fires can also start in these tunnels, and once they reach un-mined seams,

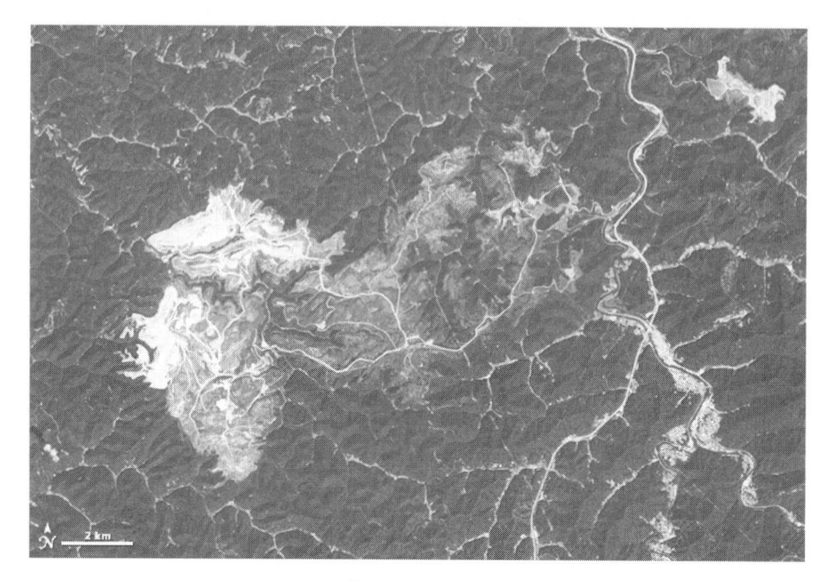


FIGURE 6.1 A strip-mined area in West Virginia. Darker areas are forest-covered, large lighter areas have been strip-mined for coal, and thin, curving white lines are either roads or stream valleys. Image from Landsat satellite provided by NASA's Global Land Cover Facility and the NASA Earth Observatory. See R. Lindsey (2015) Mountain top mining, West Virginia, NASA Earth Observatory (earthobservatory.nasa.gov).

they are nearly impossible to put out. These fires can burn for decades, leading to noxious emissions (carbon monoxide, sulfur dioxide, and mercury); potentially catastrophic collapse of roads, bridges, pipelines, and buildings; and the possibility of igniting surface fires. The US alone has had nearly 100 such fires in nine states. Pennsylvania has the largest number of currently active fires, reflecting its extensive coal deposits, long mining history, and the difficulty of extinguishing them. Perhaps the most famous is Centralia, a once-thriving town in central Pennsylvania where an underground coal fire started in 1962, when burning trash was allowed to come in contact with a coal seam. The fire spread so extensively in the next few decades that the town had to be abandoned. The fire is still burning and spreading. Centralia is now famous as a setting for science fiction and horror movies.

Many of these problems can be mitigated with best practices management. Government regulation and enforcement have an important role to play. Modern strip-mining, if done correctly, moves topsoil off to the side, removes the coal, and replaces the topsoil once mining is complete, recontouring the land to something approaching the original shape. There is minimum long-term disruption to surface drainage and groundwater if such practices are followed. Current regulations promote this, but some companies request variances or even cheat, violating the rules while hoping to gain a short-term business advantage. In my experience, most companies actually prefer clear, well-enforced regulations that lead to environmentally desirable outcomes because it levels the playing field for responsible companies and does not reward irresponsible companies that take short-cuts, cause environmental harm, and make life difficult for the industry as a whole. Company shareholders, governing boards, and bondholders have an important role to play, making sure management takes the environmental high road and does not leave the company exposed to future liability. Investors who fail to do this risk losing money. This advice applies beyond coal and land use. Consider the large losses experienced by shareholders in Volkswagen, a major automobile company, in 2015. Volkswagen managers intentionally misled consumers and regulators about the emissions and fuel economy of its diesel-powered cars. The company has lost billions of dollars in market capitalization and faces huge future liability as a result of these unethical actions.

Burning coal can lead to other problems besides land degradation. In addition to $CO_2$ emissions (Chapter 8), by-products of coal burning include heavy metals such as mercury and arsenic, as well as soot and other small particulates. The latter are especially damaging to human health, as small particles can penetrate deeply into lung tissue.

Atmospheric pollution by heavy metals such as mercury and lead is a problem because it is virtually impossible to clean up once it has spread, and it can inflict significant long-term damage to human health. It's useful to first consider the example of lead. It's better studied than mercury, but its health effects are similar – both metals are potent neurotoxins. Lead has been implicated in health problems such as anemia, hypertension, impaired concentration, hearing loss, seizures, nausea, miscarriages and stillbirths in adult females, and reduced sperm count and abnormal sperm in adult males. Lead is more dangerous for children because it affects brain development. Our problems with lead are a legacy of past industrial processes that were poorly regulated, such as manufacturing and recycling car batteries and, before the 1990's, the use of leaded gasoline, probably the largest source of lead in urban environments today. Mercury also comes from a variety of industrial sources, but coal-fired power plants are a major source.

## THE TOXIC LEGACY OF LEAD

Lead's sad history is described in excellent books by McGrayne (2001) and Bryson (2004). I summarize it here because this history has strong similarities to a number of natural hazard and environmental issues discussed in this book: initial scientific ignorance followed by increased understanding and attempts by scientists to communicate the problem to the general public and policy makers, followed by industry denial, cover ups, and even persecution of scientists attempting to describe the problem and find solutions. It also bears on our three themes: communication, long-term impacts, and economic consequences.

Ingestion of lead from food and from air leads to different health outcomes. When ingested in food, lead tends to bind with calcium and is incorporated into bone, where it is relatively harmless for adults (this is not true for children). In contrast, when we are exposed to airborne lead, it enters the bloodstream directly through our lungs, with immediate toxic effects. This is why leaded gasoline was a bad idea: It led to high levels of lead in urban air, affecting people in densely populated cities. *Time Magazine* named it one of the 50 worst things ever invented. It was finally banned in many countries between the late 1970s and the 1990s.

Blood lead concentrations in humans dropped steadily after the late 1970s, as the use of leaded gasoline declined. However, it was a long battle between the invention of this poisonous substance (known as Tetra-Ethyl Lead, or TEL) in the early 1920s by Thomas Midgley, who was working for General Motors (GM) at the time, and its final removal from the marketplace decades later. GM and Standard Oil (now Exxon) formed a new company, Ethyl Corporation, in 1924 to manufacture and market TEL, and they made Midgley a vice president. Within its first year of operation, numerous workers died of lead poisoning. Even Midgley suffered health effects from his research and must have been aware of the health problems his product would cause. Midgley died in 1944, and he may have committed suicide. Ethyl Corporation continued selling TEL and licensed the UK rights to Octel Corporation, which later became Innospec. Though leaded gasoline is now outlawed in both the US and UK, Innospec continues to export TEL to the third world, taking advantage of lax regulations to sell a product known to be hazardous to human health.

Blood lead concentrations in humans have still not returned to pre-industrial levels, partly because lead is very long-lived in the environment. Mercury pollution will have a similar longevity and will likely be present in our environment long after we stop using coal. Lead remains present in the top few centimeters of soil in many urban environments (e.g. Figure 6.2) and is easily remobilized in dust. Scientists have shown a strong correlation between the amount of lead in soil and blood lead concentrations in children, and between blood lead concentrations and various measures of neurological and intellectual development in children, including IQ, attention span, behavioral problems, and violence. This remains a poisonous legacy of industrial society's addiction to high-octane gasoline and TEL for much of the 20th Century and to a lesser extent, the widespread use of leaded paint until the 1970s. In effect, we've scored an "own goal" on our species by stunting the intellectual development of some of our children and grandchildren, especially those from low-income families who tend to live in polluted inner-city environments.

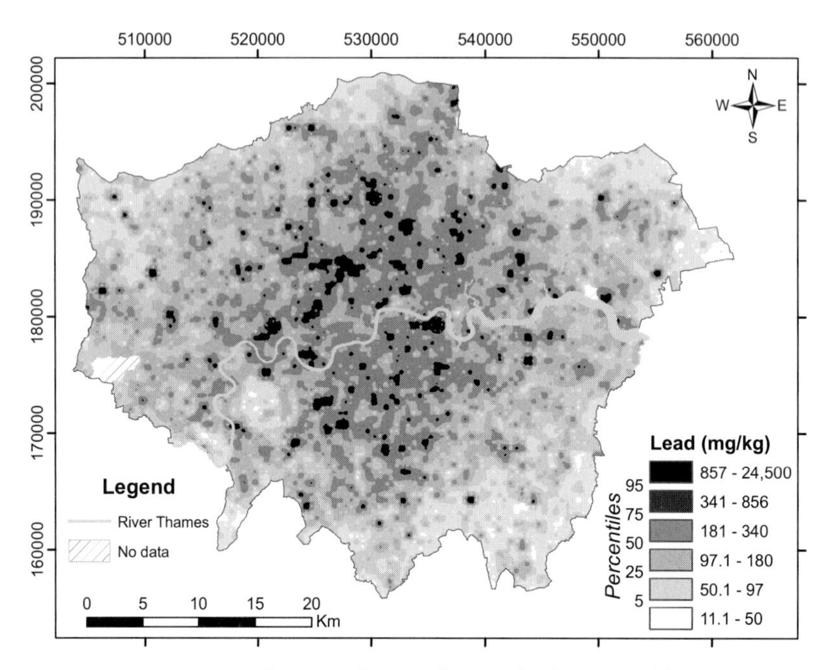


FIGURE 6.2 Lead in topsoil in London, England, measured by the British Geological Survey, 2008–2009. The scale bar on the right is in units of milligrams (mg) of lead per kilogram (kg) of soil (mg/kg). Levels above 100 mg/kg are generally considered unsafe, although the Dutch government sets a standard of 40mg/kg. Note that most of central London exceeds even the higher value. Most major urban centers show similar levels of lead contamination in their topsoil. This figure is reproduced in color in the online Appendix. Reproduced by permission of the British Geological Survey © NERC. All rights reserved. CP14/021.

My own understanding of airborne lead pollution did not come until I was a graduate student with two young children in the mid 1970s. At that time, I was struggling to understand the mysteries of mass spectrometry. This is a technique for measuring the isotopic ratio of certain elements (isotopes are chemical elements that have the same number of protons and electrons but different numbers of neutrons). I was interested in the use of various isotopes of lead for figuring out the age of certain rocks. I attended a lecture by Clair Patterson, a famous geochemist from Caltech in Pasadena,

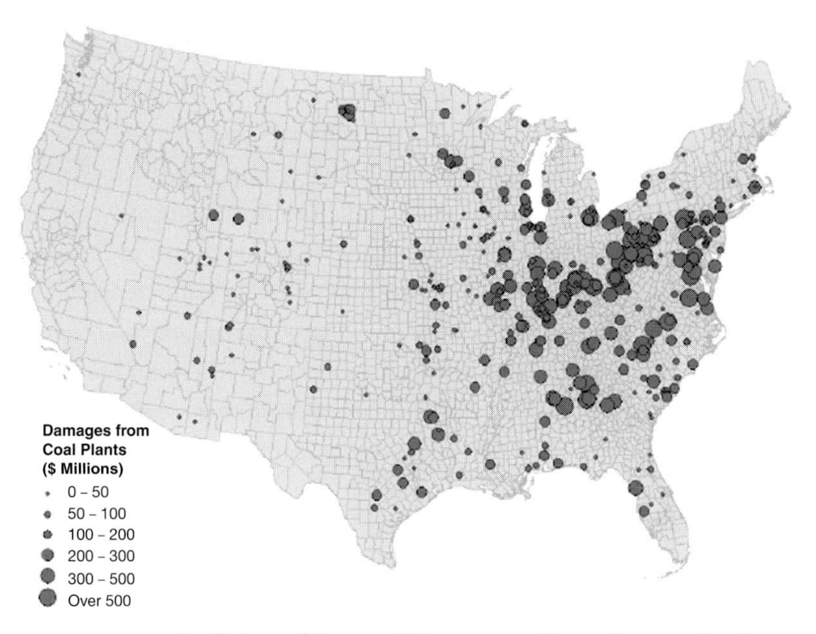


FIGURE 6.3 Estimated "external" (unaccounted) costs in 2005 of 406 coal-burning power plants in the US. These costs mainly reflect premature deaths from air pollution, based on estimated lost earning potential and taxes, and do not include "upstream" costs (e.g. land degradation from coal mining) or "downstream" costs (e.g. surface water pollution from mercury or fly ash, groundwater pollution, or global warming). From National Academy (2010).

California, hoping to learn more about the subject. Patterson had first figured out the age of the Earth using this approach (it turned out to be about 4.5 billion years old, much older than previous estimates). Patterson was a god to anyone doing geochronology – the science of measuring the ages of rocks. To my surprise, Patterson didn't talk about geochronology at all. Instead, he talked about the problem of atmospheric pollution from leaded gasoline and its harm to young children, especially in urban areas with lots of cars. My children soon began to resent the numerous lifestyle changes I ruthlessly imposed on them in the hopes of limiting their lead intake.

The behavior of industry and the media as these problems became apparent is instructive. Automobile companies predicted dire consequences for engine life, maintenance, and performance if lead was banned from gasoline. These concerns proved to be groundless. Either company engineers were not especially knowledgeable about how engines actually work (a disturbing possibility), or the companies were being untruthful, hoping to avoid the cost of redesign (also disturbing, given the damage they were causing to people's health). Ethyl Corporation, Octel, and related companies vigorously defended their main product and profit source. Scientists and medical doctors who suggested banning of lead in gasoline on grounds of public health were ridiculed as naïve idealists, clearly incapable of understanding the intricacies of automotive engineering, transportation infrastructure, and basic economics.

Patterson had turned his chemical skills to the issue of lead pollution beginning in the late 1950s. At the time, many environmental chemists had suggested that lead levels in the environment were not greatly elevated above background, hence leaded gasoline and other modern lead sources should not be a concern. Patterson used lead concentrations in Greenland ice cores and other pre-industrial samples to show that atmospheric lead pollution had actually increased by huge amounts since the introduction of leaded gasoline; previous studies had used contaminated samples for background indicators, and thus had greatly underestimated the problem, an indication of how bad the problem actually was. In ways analogous to how sediment layers preserve information about earthquakes and tsunamis (Figure 4.1), the ice in Greenland preserves a record of most things in the atmosphere, including lead. Ice has the added benefit that, like trees, you can count the annual "rings" (snow layers) to determine its age. Earlier studies of lead in the environment had claimed no significant recent increase in lead because they had failed to go back far enough (to unpolluted earlier times). Like the short earthquake record in northern Japan (Chapter 4), this is another example of the need for a long time span to get a representative record.

Ethyl Corporation, through its industry lobbying organization, the American Petroleum Institute (API), attempted to discredit Patterson's work. In a *Wall Street Journal* article published September 9, 1965, API stated that the while his findings "may be of academic interest ... they have no real bearing on the public health aspects of lead. Contrary to Mr. Patterson's conclusion, the mass of evidence proves unquestionably that lead isn't a significant factor in air pollution and represents no public health problem in any way." Given the well-known toxicity of lead and Patterson's compelling evidence for its planet-wide dispersal, this was an astonishing statement and, unfortunately, went unchallenged in the *Wall Street Journal* article. The *Journal* further undermined Patterson's message by quoting a medical professional, Dr. Leonard Goldwater at Columbia University, who characterized Patterson's statements as "misleading." Goldwater had studied blood lead concentrations in humans over a 30 year period, a time span when there wasn't much change. Goldwater's mistake was similar to the one made by TEPCO managers at Fukushima – failure to think long term (even Goldwater's oldest samples were contaminated with lead). Dr. Herbert Needleman, a scientist at the University of Pittsburg who studied the neurological effects of low levels of lead in children a few years later and showed that even small amounts of lead would hamper brain development and reduce intelligence in children, was similarly targeted by industry and special interest groups, with some accusing him of research misconduct. These charges were eventually refuted, but at considerable cost to Needleman. Again, the media were unduly compliant in airing the views of special interest groups, at the expense of Needleman's research reputation, without serious investigation of the allegations. To my knowledge, API never apologized for its role in delaying much-needed industry changes and the resulting damage to public health.

The pattern of shooting the messenger is eerily familiar. Scientists who later fought against manufacturers of chlorofluorocarbons because of their impact on stratospheric ozone or against Big Tobacco because of smoking's impact on lung disease all faced similar obstacles from a hostile industry and special interests, often aided by a naïve and easily manipulated media. Scientists studying carbon emissions and links to global warming face similar obstacles. Perhaps not surprisingly, API is again involved (see Chapter 8).

Journalists would do well to remember the line quoted at the beginning of this chapter ("Follow the money") when reporting the views of special interest groups. Dr. Naomi Oreskes, a professor of history at the University of California, and Dr. Eric Conway, a historian at NASA's Jet Propulsion Lab, describe in their book *Merchants of Doubt* the process by which well-funded special interest groups, a few fringe scientists, and a compliant, unquestioning media are able to delay needed change, even in the face of overwhelming scientific evidence. Their book should be required reading for every journalism student.

## MERCURY AND COAL

Given that coal has been burned in large quantities on every continent on Earth except Antarctica for nearly two hundred years, it should come as no surprise that heavy metal pollution from coal, including mercury, is widespread and may be as much of a problem as lead. But how widespread? And what is the relative influence of coal versus other sources? Mercury is also used in industrial processes such as PVC production, where it is used as a catalyst, and in small-scale gold mining. In some coastal areas, concentrations are high enough that they can have devastating consequences for marine life, especially higher order life with complex neurological pathways. In parts of southeast Florida, dolphins have had their immune systems suppressed by mercury exposure, making them susceptible to a variety of diseases. For those of us living in Florida, dolphins are one of several important indicator species (our canaries in the coal mine) for ecosystem health. Unfortunately, state spending on environmental monitoring has been slashed in the last decade.

In the open ocean, mercury is present in much lower concentrations compared to polluted freshwater bodies and coastal environments and, hence, more difficult to analyze. At the risk of mixing metaphors, a good canary in the coal mine for mercury, and coal's role in its dispersal, is the amount of mercury found in tuna and swordfish. Both are pelagic (deep water) fish that spend most of their lives far from coastlines and far from local sources of pollution. Nevertheless, most modern tuna and swordfish, regardless of where they are caught, have measurable amounts of mercury in their flesh. Most of this comes from coal. Coal is the only major source of mercury capable of lofting large amounts of this element to high altitudes (via smokestacks), promoting widespread dispersal. Coal has also been used as a major energy source throughout the world for more than 200 years, so there has been ample opportunity to spread its trace pollutants throughout the global ocean. According to Streets et al. (2011), coal-fired power plants were the major source of mercury pollution during most of the 20th century, although gold mining may now be taking over this dubious distinction.

Tuna and swordfish are especially good indicators of mercury pollution in the deep ocean because these fish spend a lot of time there and are near the top of the food chain (humans sit at the top since we eat tuna and swordfish). These predator fish tend to concentrate small amounts of mercury pollution in their tissues, allowing it to be easily measured. One way to think about this is that small organisms in the ocean have small amounts of mercury in their tissues, not much higher than the concentration in the water itself. But fish that eat these organisms over several years will accumulate some of that mercury in their tissues. When a big predator fish comes along and eats up this smaller fish, it also gets a good dose of mercury with its meal. After several years of eating nothing but smaller fish with mild doses of mercury, you get a big fish with lots of mercury. The process is called bio-accumulation. Predator fish at the top of the food chain can have mercury concentrations in their flesh that are millions of times higher than ambient seawater (Mason et al., 2012).

Table 6.1 *Methyl mercury in fish (ppm)**

| Seafood Type | Average | Maximum |
|---|---|---|
| Anchovies | 0.04 | 0.34 |
| Cod | 0.11 | 0.42 |
| Lobster (Spiny) | 0.09 | 0.27 |
| Lobster (Northern) | 0.31 | 1.31 |
| Salmon | 0.01 | 0.19 |
| Sardines | 0.02 | 0.04 |
| Shark | 0.99 | 4.54 |
| Swordfish | 0.97 | 3.22 |
| Tilefish (Atlantic) | 0.15 | 0.53 |
| Tilefish (Gulf of Mexico) | 1.45 | 3.73 |
| Tuna (albacore, canned) | 0.35 | 0.85 |
| Tuna (fresh/frozen) | 0.38 | 1.30 |

* Summarized from Table 2–13 in Nesheim and Yaktine (2007).

Mercury is extremely toxic to humans, even more so than lead. It can be a bit confusing to talk about this quantitatively, because mercury in the environment and in food comes in several forms, with varying levels of toxicity (inorganic mercury is less toxic than methyl mercury, a type of organic compound). Also, exposure limits change with time scale and the nature of exposure – it makes a difference whether you're talking about a one-time exposure through food or long-term exposure to vapors (in the latter case, even small doses can be hazardous). To give some idea of the tiny amounts involved, the US Food and Drug Administration sets a limit for food of 1 ppm (part per million) for methyl mercury. Most fish high on the food chain, such as tuna, shark, swordfish, and tilefish, come close to this limit or occasionally exceed it (Table 6.1).

The mercury standard for drinking water is even lower: The US Environmental Protection Agency sets a limit for inorganic mercury in drinking water of 2 ppb (two parts per billion). These are extremely small amounts that are difficult to measure except with specialized equipment, but it's a good indicator of the extreme toxicity of this substance. Groundwater and surface waters downwind of some coal-fired power plants routinely fail to meet the drinking water standard.

Mercury's toxicity has been known for a long time. The Mad Hatter character in *Alice in Wonderland* is based on hat makers in 19th-century Europe who used mercury to cure the felt used in certain hats and often exhibited neurological symptoms of mercury poisoning. Dr. Jane Hightower, a physician in San Francisco, California, has written a book discussing the problems of mercury exposure and how to avoid them. One of her common sense suggestions: Eat less tuna and other fish near the top of the food chain (humans can bio-accumulate too!).

A stark illustration of mercury's power to poison has become known as Minamata disease, named after the Japanese fishing village where the disease was first discovered in 1956. Several thousand people were diagnosed with the disease, and at least 1,700 people died, a very high fatality rate. The disease initially damages vision, hearing, and speech, and as it progresses, it leads to paralysis, insanity and death. Many children in the region developed cerebral palsy, a crippling disease. By 1959, the disease was firmly linked to mercury poisoning. A nearby chemical factory that used mercury in their industrial process had not invested in a mercury recovery system; waste products containing mercury were dumped directly into the nearby bay. The mercury bio-accumulated in fish and shellfish that were eaten by the victims. Pregnant women exposed their fetuses to toxic levels of mercury, explaining the high incidence of childhood cerebral palsy. In a series of events that is sadly familiar, the chemical company responsible for the pollution tried to discredit researchers from nearby Kumamoto University who were investigating the tragedy and whose research results pointed unequivocally to the company's own waste products as the culprit. Unfortunately, many companies, when faced with this type of pollution or health issue, follow a similar strategy of denial and cover up. In my opinion, a better long-term approach is to admit the problem, work with scientists and engineers to find a solution, and then invest in the fix. While more expensive in the short term, this would reduce future legal liability and ensure a market for your product – killing or sickening future customers is bad for pubic relations and is a poor long-term business strategy.

Since mercury is extremely toxic, especially for pregnant women and small children, and since some seafood can have pretty high concentrations of mercury, often above levels generally considered safe to eat except in small quantities, most toxicologists recommend limiting consumption of fish high on the food chain (Table 6.1). The seafood industry has resisted labels warning that their product may be unsafe for children and pregnant women due to mercury contamination. One argument the industry makes is that any mercury, if present, may be natural (volcanoes emit small amounts of mercury). In my opinion, this is a specious argument. First, it does not really matter where the mercury comes from – it's all toxic. Remember, the mercury from coal is "natural"; the only thing that is not natural is its recent remobilization into the modern environment. Second, many studies show that the great majority of mercury or methyl mercury in the environment does in fact come from industrial sources, including coal combustion. Rather than sweep the problem under the rug, I'd like to see the seafood industry step up and do its own mercury testing, publish the results, and work with environmental groups to regulate and reduce mercury pollution throughout society. The industry could become a powerful force for good and help to ensure the health of its customers, the long-term health of the world's ocean, and its own long-term survival.

Some batteries contain mercury. When thrown away, they can wind up in incinerators, where the mercury is dispersed into the atmosphere, or in landfills where it can contaminate groundwater. Putting a small monetary deposit on mercury-bearing batteries to encourage recycling would be a simple way to reduce pollution from this source. Cremation facilities are another source of mercury pollution – many people (including me) have fillings in their teeth that are an amalgam (mixture) of silver and mercury. It's quite safe in this form, but when mobilized at high temperature, it enters the atmosphere, drifts downwind, and then rains out into the soil. We shouldn't have to wait for legislation from politicians – manufacturers and retailers of mercury-bearing products could come up with their own deposit and recycling scheme. Cremation facilities could likewise remove the amalgam-filled teeth of their clients prior to cremation. Both businesses could probably boost sales by advertising their "green" credentials.

Mercury is a naturally occurring element, present in many rocks in low concentration and emitted in low concentrations by many volcanoes. One question the reader might ask is how do we know mercury in the ocean didn't come from volcanoes? A second question might be, if mercury is naturally occurring, why have we not evolved some resistance to it, and why is it so toxic? The answer to the first question has to do with chemistry and the formation of stable compounds. Mercury binds with other elements to make relatively inert substances that geologists call minerals. These are chemical compounds (groups of elements) that form stable crystal structures and exist naturally. Once bound up in such a crystal structure, mercury is much less likely to cause problems. A common mercury-bearing mineral is called cinnabar – it's a beautiful dark red mineral, similar to cinnamon in color and made up of equal parts of sulfur (chemical formula S) and mercury (chemical formula Hg). (Cinnabar's chemical formula is HgS.) Volcanoes emit lots of sulfur, so most of the mercury atoms emitted by the volcano quickly bind with sulfur and are deposited nearby as cinnabar. Cinnabar is one of the principle ores of mercury, and it should not come as a surprise that most of it is found associated with ancient volcanoes or hydrothermal systems (hot springs are usually associated with volcanoes). As long as we don't try to break up mercury-bearing minerals such as cinnabar, for example by digging them up and heating them to release the mercury, they are quite safe in their natural form.

Another clue that most mercury in the modern ocean is anthropogenic (in other words, it comes from our own industrial activities, not from volcanoes) comes from the fact that the concentration of mercury in the ocean has increased greatly over time. Exactly how much is an important question. Shannon Palus, a freelance science writer, has put the question succinctly: To what extent are we poisoning ourselves? It is generally agreed that the increase has been at least a factor of three since the Industrial Revolution. A recent estimate based on modeling suggests an increase by factors of 5 to 6 since the mid 15th century. Perhaps the best estimate of mercury's recent increase is based on a study reminiscent of the trenching studies done by geologists to decipher earthquake history, as described in Chapters 4 and 5. Dr. L. Q. Xu and colleagues at the University of Science and Technology of China studied the eggshells of marine seabirds on a small island in the South China Sea. Just as fish at or near the top of the food chain concentrate mercury in their tissues, seabirds that eat the fish also concentrate mercury, and it winds up in their eggshells. Xu and colleagues dug a pit on the island and carefully collected and catalogued eggshell fragments from different depths (remember superposition – deeper ones are older). By dating the shell fragments and analyzing their mercury content, the researchers determined how mercury concentrations have changed in the ocean over the last 700 years (that age being the oldest shell fragments at the bottom of the pit). It turns out that mercury has increased tenfold over the last 700 years, with most of the increase since 1850, i.e. since the beginning of the Industrial Revolution and the widespread use of coal.

The answer to the second question (our lack of natural resistance to mercury), like most toxicity issues, has to do with concentration. Until the Industrial Revolution, the amount of mercury that most people were exposed during a lifetime would be extremely low (mad hatters aside). Much like lead, it is the recent additions to the environment that are causing problems.

Now it's time to explain how mercury got into coal in the first place. To understand the process, let's first consider the Brita filter. In 1966, a German company named Brita began selling water filtration products. Today, the Brita filter is one of the top selling portable water filters in Western consumer markets. The Brita filter includes activated charcoal (essentially pure carbon) made up of small particles of charcoal formed by burning a woody product such as crushed coconut shell in a partial vacuum (otherwise the carbon in the wood oxidizes, forming $CO_2$). Charcoal is a porous material, which means it has a high surface area and lots of little holes. Larger ions and molecules in the water such as lead, mercury and various chlorine compounds are adsorbed (stuck) onto the charcoal's surface. If you're worried about lead in your drinking water from the solder commonly used in plumbing or you don't like the taste of chlorine, the Brita filter is a pretty good option.

Coal occurs in sedimentary layers formed over millions of years. It represents the accumulated organic remains of plants that grew in marshes and swamps. As the marshes were slowly buried, the plant compounds, composed of carbon, nitrogen, oxygen, water, and other elements and compounds, were gradually reduced in volume and pressed together, with the nitrogen, oxygen, and water being driven off by heat and pressure. Eventually a layer of pure carbon (coal) forms within the sediment. Given this origin, you can understand why coal deposits often contain plant fossils (the fossils are how geologists first figured out how coal forms).

Occasionally, the coal layers can act like a Brita filter. Hydrothermal waters emanating from great depth or groundwater percolating from above may have picked up mercury. As these waters pass through the coal layers, mercury is trapped, leaving purified water but contaminating the coal. More often, it is the original plants themselves that trapped the mercury. During their lifetime, plants cycle large amounts of water through their systems. Mercury tends to get trapped in roots and woody plant tissues during this cycling, just as a Brita filter traps contaminants.

Perhaps this or similar process contributed to our low tolerance for mercury. We evolved at Earth's surface in a virtually mercury-free environment because coal and woody plant tissues had trapped and sequestered most of the mercury. If so, the last thing we'd want to do is dig up a fossilized mercury trap (coal) and burn it, putting the mercury back into circulation. But that's exactly what we do when we use coal for power, just like Midgely, GM, and Standard Oil did with leaded gasoline. We can fix this if we're smart enough to deal properly with the waste products of coal combustion, which is technically possible (see Carbon Capture and Storage, below) as well as the waste products of other industrial processes that use mercury.

What is the cost to the global economy of mercury and other pollution from coal? While it's hard to pick out the costs associated with any one pollutant, it is well known that the combination of heavy metal pollution, sulfuric acid pollution, and pollution from soot and other small particles means that coal-fired power plants can be a major contributor to low air quality and poor health. Adverse health effects include lung cancer, emphysema, bronchitis, asthma, and heart attack. The proportion of polluted air contributed by coal compared to other sources has also been studied extensively. This allows us to quantify the health effects and associated risks and costs of burning coal, even if we can't single out the ill effects of mercury.

## DEATHS FROM COAL VERSUS DEATHS FROM NUCLEAR POWER

A report by the US National Academies of Sciences, Engineering and Medicine published in 2010 quantified environmental impacts in the US from burning coal, painting an unflattering picture. Their conclusions probably apply to any industrialized country that burns a lot of coal.

The adverse effects of air pollution from coal are called "externalities" by economists, as they are costs not accounted for by a company's internal cost assessments. In my opinion, this represents a failure in the regulatory structure, since other industries, at least in the industrialized West, are required to pay directly for disposal of their waste products and all related costs. Instead, for coal, these costs are borne by society as a whole. John Hofmeister, in his book *Why We Hate the Oil Companies*, points out that the oil industry is one of a small number of industries able to dump its main waste product ($CO_2$) free of charge. Much the same can be said about coal-fired power plants, and we could add the costs of mercury, sulfur, soot, and other particulates that come from these plants, especially older ones not equipped with the latest pollution control devices.

When emissions from coal burning are poorly regulated, it also represents a failure by the financial accounting industry, in particular failure to consider longer-term financial implications. Air pollution from a poorly regulated coal-fired power plant represents a huge potential future liability. At some point, a citizen whose nonsmoking spouse, parent, sibling, or child died from emphysema or lung cancer is going to sue the utility operating such a plant, much as smokers or their families started to sue tobacco companies in the 1980s. Initially, the tobacco suits were unsuccessful; tobacco companies, with the help of an uncritical media, were able to discredit scientific studies linking smoking and lung cancer. But eventually the evidence became overwhelming, common sense prevailed, and plaintiffs started to succeed. The sheer number and potential future costs of additional lawsuits eventually became too large for the industry to ignore, resulting in large settlements in favor of the plaintiffs and laws that limited the ability of "Big Tobacco" to market its addictive and unhealthy products, at least in Western countries (like leaded gasoline and coal, the primary "growth" market for tobacco is now in poorly regulated third world countries). If I were an executive at a power utility that operated a coal-fired power plant, I would want to insulate the company from similar future liabilities by making sure that the latest pollution control devices were installed and operating properly. Hopefully these executives will not dodge the bullet by moving operations to third world countries.

To quantify the "externalities" associated with coal, let's look at the 2010 National Academy report in more detail. The committee used data for 2005 and calculated that $62 billion (US) in excess or external costs were associated with air pollution from coal-fired electric plants that year (p. 340 of the report). This number exceeds the cost of the other major source of air pollution in the US – transportation. Of the $62 billion total external cost, 94 percent represents premature mortality (p. 94). In other words, what economists and financial accountants call an externality, most other people would call death. The $62 billion cost estimate assumed that an average life, in statistical terms, was worth $6 million based on lost earnings potential and taxes (p. 85), implying that the models used in the report predicted approximately 9,700 premature deaths per year associated with production of electricity from coal ($62 billion times 94 percent divided by $6 million). Note that this cost (in dollars or in lives) does not count "upstream" costs, such as those associated with strip-mining or water pollution, nor any "downstream" costs, such as those related to waste disposal (large volumes of sludge and fly ash, both contaminated with heavy metal, or costs associated with global warming) or coal fires. The estimate of the societal cost of burning coal should therefore be considered a minimum estimate.

The report also noted that a large amount of air pollution was caused by a relatively small number of older coal-fired plants. In other words, the majority of coal-fired power plants operating in the US today are actually pretty clean. Technology exists to clean up the older, dirtier plants, but the owners of those plants have decided it's a poor investment, and weak regulation and enforcement, plus political connections, allow them to stay in business. Perhaps it will be those plants and the companies that own them that will be the focus of future lawsuits by citizens whose loved ones die as a result of easily preventable air pollution.

While the 2010 National Academy report represents a major milestone in efforts to understand the full range of societal damages caused by burning coal, it did not seem to have much impact on the public discourse. Environmentalists were not out picketing electric utilities the day after its release, and no new legislation was passed the following year. I suspect one reason for this is the writing style of the report, typical of many scientific studies. At 506 pages, it was quite a heavy read. Even the Executive Summary ran on for 19 pages, far too long in my opinion and a clear violation of Strunk and White's Rule #1, *Use fewer words*. Moreover, the Executive Summary, which is all that most people would have time to read, did not even list the mortality statistics directly. Instead, the key information was buried in the body of the report, and an astute reader who wanted to know how many people were killed each year by coal pollution would have quite a job teasing out this statistic. As with many reports written by committee, clarity was sacrificed for consensus. Perhaps the committee thought it too controversial to state clearly that due to a relatively small number of poorly equipped power plants, coal burning leads to the premature deaths of thousands of people in the US each year due to air pollution. But in my opinion, that is exactly what they should have said in the first sentence of the first paragraph of the Executive Summary. A list of the poorly equipped plants causing most of the pollution and fatalities would also have been useful.

For comparison to the estimate of 9,700 premature deaths per year in the US from coal, let's now consider the number of premature deaths from nuclear power. It's difficult to come up with a precise figure. For example, how would you account for possible future deaths due to low levels of radiation from normally operating plants, since background natural radiation is at a similar level? One way is to look at extreme events and consider the resulting estimate an upper bound. Chernobyl was the world's worst accident related to nuclear power. It occurred in Ukraine in the former Soviet Union in 1986. The core meltdown and subsequent explosions at Chernobyl exposed millions of people in the former Soviet Union and Eastern

Europe to radiation (the reactor lacked a containment vessel, making the local populace vulnerable to even minor accidents). A 2005 study by the International Atomic Energy Agency (IAEA) estimated 4,000 additional deaths as a result of that accident. This number was challenged in a subsequent 2006 Greenpeace report. The latter study compiled reports from a number of Russian and Ukrainian studies, many of them unpublished in the Western scientific literature, and suggested that up to the year 2004, approximately 200,000 additional deaths had occurred in Russia, Ukraine, and Belarus, the three countries most affected by Chernobyl radiation. In 2008, a UN report issued a more authoritative study and estimated a total of 9,340 additional deaths in all countries from Chernobyl (UNSCEAR, 2008, Table D-24, p. 197, summing all sources of "Predicted Excess Cancer Mortality"). Although the much larger estimate from Greenpeace has not been peer-reviewed (and I suspect is incorrect), let's accept it for the sake of argument (it will make our subsequent calculation a maximum value). Let's further assume that over the same time period (1986–2004), an additional 100,000 deaths occurred worldwide from nuclear power, from Chernobyl-related deaths in other countries, and from unreported low-level radiation incidents at other power plants (I know of no credible study that suggests anything near this level of mortality from either source, but again this will make our estimate a maximum value). If 300,000 deaths had occurred over 18 years (1986–2004), this would imply approximately 16,700 deaths per year worldwide from nuclear power. Since the US population represents roughly 5 percent of the global total, this in turn implies that about 835 deaths per year in the US are associated with nuclear power. Recall that this spans an 18-year period that includes the worst nuclear accident in history, at a nuclear reactor lacking a containment vessel, and using mortality figures that are almost certainly inflated. Ever since Chernobyl, all nuclear reactors have been fitted with containment vessels (most reactors had them before the accident). If we use the more plausible numbers from the 2008

UNSCEAR estimate, we obtain fewer than 20 premature deaths per year in the US.

Now we can finally compare nuclear and coal directly, at least for the US: a maximum of 835 (or more likely, 20) nuclear power-related deaths per year, versus 9,700 deaths per year from coal. In terms of relative risk, nuclear wins hands down over coal. This point is not widely appreciated by the public, the media, policy makers, and many environmentalists.

It's also worth noting that 9,700 premature deaths per year in the US from coal is still a relatively low number given the amount of power produced. In a country of approximately 300 million, it represents an annual fatality rate of less than 0.003 percent. For comparison, the US experiences more than 30,000 deaths per year from automobile accidents and similar fatality levels from gun-related homicides, suicides, and accidents.

Failure to appreciate relative risk and cost versus benefit permeates many aspects of modern society. Let's look at a few other examples. In the US, vast sums of money have been spent in response to the terrorist attacks of September 11, 2001, which killed approximately 3,000 people. These investments have clearly made the country safer, but government efforts would likely fail a rigorous cost-benefit analysis. Consider that approximately 30,000 Americans are killed every year by guns from a combination of homicides, suicides, and accidental shootings. Most homicide victims know their killer. In a statistical sense, Americans have more to fear from friends, neighbors, and family members than from foreign terrorists.

Afghanistan is fighting a nasty insurgency by the Taliban, a group of religious extremists. Western countries have spent hundreds of billions of dollars in the last decade to help the country defeat the Taliban and develop a modern democracy, with some success. In 2007, a total of 1,400 Afghan civilians died as a result of the insurgency, from all sources including bombings and other terrorist attacks, as well as enemy and friendly fire. During the same period, according to a report by the Asian Development Bank, approximately 2,300 civilians died in Kabul, the capital, from air pollution, in this case mainly caused by cars, dust from unpaved roads, and cooking fires. An investment of less than $100 million in cleaner cooking stoves, paved roads, and better public transportation would probably cut the number of fatalities from air pollution in half, at far less cost than military expenditures to reduce civilian fatalities from Taliban attack by the same amount.

In discussing relative risk, it should be emphasized that air pollution is one of our largest killers. Lomburg (2014) estimates that indoor air pollution from cooking fires kills approximately three million people per year, mostly in the third world. The European Environmental Agency estimates that air pollution, mainly from coal-fired power plants and diesel fuel burned by cars and trucks, kills 400,000 Europeans each year. Perhaps not surprisingly, diesel manufacturers have tended to downplay or hide the health risks associated with their product.

The World Health Organization estimated that a total of seven million people died in 2012 from all types of air pollution. This means that air pollution, mainly from the use of carbon-based fuels for electricity, cooking, heating, and transportation, is by far our largest killer, eclipsing deaths from all natural hazards, chemical spills, nuclear radiation, terrorist attacks, war, and other catastrophes. Emphasizing this immediate risk, rather than the long-term problem of global warming, might be a better strategy for environmentalists and policy makers seeking to reduce fossil fuel use.

Given the number of fatalities from air pollution, you would think governments would do everything possible to discourage use of fossil fuels. If so, you would be wrong. Many countries actually subsidize fossil fuel use. The International Energy Agency estimates that direct global subsidies for fossil fuel totaled $490 billion in 2014. In 2015, the International Monetary Fund (IMF) published a broader study, concluding that the global cost of fossil fuel subsidies is $5.3 trillion per year, roughly 6 percent of global GDP. Most of the subsidies go to coal and oil. The biggest offenders are China, the US, Russia, India, and Japan, but many countries follow this dubious practice. While some of the cost represents foregone taxes, exploration incentives, or long-term costs associated with global warming (the latter number is difficult to estimate), most of the costs involve local pollution, health effects, and premature death. In effect, countries are paying to kill their own citizens.

## CARBON CAPTURE AND STORAGE

One promising technology that has not yet been widely used to clean up coal, but in my opinion could be, is carbon capture and storage, CCS (full disclosure: I have received funding from the US Department of Energy to investigate some aspects of CCS). The CCS process involves capturing some or most of the carbon dioxide and other pollutants emitted by a coal-burning power plant or other point source of $CO_2$, and pumping them deep underground into a geological reservoir where it will (hopefully) stay for millennia or longer. Originally developed as a way to reduce the impact of coal on global warming, it has the nice side benefit that many other pollutants, including mercury, sulfur, and particulates, are greatly reduced as well, pumped deep underground where they can't do any harm. While it does increase the cost of power generation by 20–30 percent, it greatly reduces the overall cost to society via long-term health benefits. Unfortunately, the great majority of utilities that operate coal-fired power plants have not yet adopted this technology because current regulations allow them to dump most of their waste products into the atmosphere free of charge. Companies that invested heavily in CCS technology would be at a competitive disadvantage, at least in the short term. Some changes in the current regulatory regime and some bolder long-term thinking in the private sector are clearly needed. The US Department of Energy, the Southern Company, and Mississippi Power are collaborating to build a state of the art coal-fired power plant in Kemper County, Mississippi. This 580 MW plant, scheduled to be operational in 2016–2017, uses coal gasification and carbon capture technology to capture and store approximately 65 percent of the $CO_2$ generated by coal combustion. In the process of capturing $CO_2$, most other pollutants are also greatly reduced. This plant shows that is possible to do better with coal. Hopefully it can be a model for the industry.

CCS also illustrates why electric cars can be better for the environment than cars powered directly by fossil fuels, even if the electricity is generated by coal. It is not economically feasible to capture $CO_2$ emitted by gasoline- or diesel-powered cars. However, the emissions from a centralized coal or other fossil fuel-fired power plant can be captured – the economics of scale work in our favor. Other emissions as well are also easily captured at big facilities, which means that electric cars have the advantage in terms of their overall air pollution footprint. Improvements in technology, stricter regulations, and public attention to the health effects of air pollution may eventually lead to widespread adoption of electric vehicles in place of gasoline- or diesel-powered vehicles.

To sum up, any rational assessment of power generation options has to include a full assessment of costs and risks associated with each option. When this full accounting is done, most existing coal-fired power plants are at or close to the bottom because of their large negative impact on the environment from air pollution. In the US, cleaning up or closing the dirtiest 15–20 percent of coal-fired power plants would go a long way to cleaning our air and making coal a more benign power source. More aggressive use of CCS technology would also help. Nuclear energy, despite the risks of radiation leakage, should be considered a viable option for the next few decades, until renewable sources of energy become more cost effective.

## WHAT ABOUT SOLAR?

When I teach a class in environmental science, students often ask why we don't simply abandon both coal and nuclear and go with renewable energy, such as solar electric power. This seems like an ideal source: Photons from the sun cause an electric current to flow in a solar panel.

Sunshine is free, so all you have to do is buy a solar panel and give up some roof space. A recent article in my hometown newspaper, the *Tampa Bay Times*, illustrates some of the issues. The article described a local business that had fully converted to solar power. At a cost of $47,000, well beyond the reach of most homeowners, the business saved approximately $2,500 per year in energy costs. At this rate, it would take nearly 19 years to pay off the cost of the investment, assuming energy costs stayed constant. This would not be a reasonable investment for most businesses, where payoff times for capital investments are typically planned to be less than ten years. However, the investment was made more attractive by virtue of state and federal subsidies that totaled $34,000. With a net cost of only $13,000, the cost of the solar electric installation would be paid off in about five years, a good investment for this business. But without the subsidies, it would not have been economic. Energy costs will have to rise a lot higher, either through scarcity, environmental regulation, or taxes; or solar electric power will have to become much cheaper, before such investments make sense on a large scale. While the cost of solar panels is dropping, most of the $47,000 cost in the above example can be attributed to activities associated with system design, permit applications, installation, and integration with existing electrical facilities. Falling costs for solar panels will not affect these other costs.

Larger, more centralized facilities have an advantage, as the high fixed costs can be spread over a larger power system. Warehouses and other large buildings with lots of empty roof space seem like ideal places. Another recent article in my local newspaper described a beer distribution company in central Florida that is spending $2.6 million to install a 1.5 Megawatt solar electric system on their roof. The system will pay for itself in about six years, after considering a federal tax credit.

Future technology improvements will also help, especially in the area of efficiency. The reason is that for a fixed roof space, many installations at present don't generate enough electricity for homeowner or business needs, making it necessary to pay for backup power. Limited efficiency also limits excess power that can be generated and sold back to the grid. Most present-generation silicon-based solar cells are quite inefficient. Although they have theoretical efficiencies in the range of 5–15 percent (i.e. they convert 5–15 percent of the energy of incoming photons into electrical current), in practice installed efficiency is often much less – the angle of a panel toward the sun may not be optimal, the panel surface tends to get covered with dust, and electrical components degrade over time. Actual efficiencies may be only a few percent. The net effect is an expensive installation that doesn't generate much electricity.

Improved installation and maintenance practices are beginning to improve efficiencies. New materials, based on the mineral perovskite, also promise higher efficiencies, in the 20–25 percent range. Perovskite is dark, red-brown mineral (chemical formula $CaTiO_3$). It was first discovered in the Ural Mountains in the early 19th century and was named after Russian mineralogist Lev Perovski. It turns out that many materials have the same basic structure as perovskite, and some artificial materials with this structure are very good at generating electricity when exposed to sunlight. These new materials can be made cheaply and can also be transparent, implying that windows themselves could one day become solar electric generators. Perovskite structures can also be "tuned" to be sensitive to different wavelengths of light. Combined with transparency, this implies that multilayer perovskite panels could be designed to exploit a broader range of solar wavelengths, not just the visible range used by existing panels.

Despite the current high costs, some countries are aggressively pursuing solar. Germany has installed a large amount of solar electric power, even though it is a northern country and not favorably situated for efficient use of solar cells. The Canadian province of Ontario closed its last coal-fired power plant in 2014 and is making up the difference with nuclear and natural gas-fired power plants, and subsidized wind and solar installations.

New start-up companies are also providing innovations that challenge conventional ideas of power generation. However, these companies face significant barriers. Many countries do not have a level playing field for energy. In Europe, "national champions" tend to get favored treatment over new entrants to the energy field despite decades of EU-related economic liberalization. In the US, some states maintain monopolistic laws that favor conventional energy sources over renewable sources such as wind or solar. In my own state of Florida, it is actually illegal for a new wind or solar energy company to sell electricity directly to the consumer at market price, or for smaller installations to feed unused energy back to the grid and be reimbursed at market value. Only a few state-sanctioned monopolies are allowed to do this. The monopolies tend to make generous donations to the election campaigns of state politicians, who in turn tend to enforce the status quo. In the last election cycle, these companies donated more than $3 million to state legislators. Not surprisingly, Florida, which has obvious natural advantages in terms of solar, lags behind much of the nation and the world in developing this obvious source of power.

Wind power and large-scale solar thermal power, where the sun is used to boil water or other fluid to drive a turbine (Figure 6.4), are also starting to play a role in the renewable energy arena. But each of these is still more expensive than coal and may require subsidy or a tax on carbon emissions or other harmful pollutants to level the environmental playing field.

Energy costs are almost certain to rise in the long run, but the amount depends on two variables: future scarcity and future environmental regulation. Right now, there is little or no environmental penalty paid in our society to burn coal, oil, or natural gas, so prices are determined mainly supply (related to resource scarcity) and demand. Some environmentalists have pinned their hopes on imminent scarcity of these fuels and consequent price rises to force conservation and increase the use of renewable energy. One clue used by economists to determine whether we might be approaching such

FIGURE 6.4 A solar-thermal plant in southeastern California. Mirrors on the ground focus sunlight on the central tower to heat a working fluid, which drives a turbine. Photo by the author.

a limit would be when oil prices become "inelastic," i.e. price increases do not lead to corresponding supply increases because of physical scarcity.

Experts have predicted for years that world oil supply will soon peak and then decline, despite rising demand (Deffeyes, 2001; Kunstler, 2006; Murray, 2013). Murray (2013) argues that the inelastic nature of oil prices in the mid 2000s proves that we are already at or close to that peak. I disagree. Murray and others failed to consider the concept of time lag and how it affects commodity prices. The Appendix includes a student exercise related to the issue of time lag and its contribution to "boom and bust" cycles in commodities like oil. Briefly, major hydrocarbon shortages are unlikely before 2050 because as oil prices rise, deposits once considered too expensive will become feasible to exploit. New technologies may be required, and they may take a few years to develop, leading to big price swings. But the resulting shortages are temporary. Over time, these newly available deposits flatten out the cost curve, limiting future price increases. Canada's oil sands represent a huge, barely tapped hydrocarbon source that is now profitable to exploit whenever the price of oil is above ~ $70 per barrel. The break-even point is falling all the time. Hydraulic fracturing ("fracking") for natural gas – where water, sand, and chemicals are injected into gas-rich shale – and a related process for oil also become economic as energy prices rise, increasing supply and moderating price increases. Sometimes the absolute price declines – in the US, fracking has caused natural gas prices to plummet in the last decade. Unfortunately, these newer energy sources tend to generate more carbon dioxide during extraction. One day, new deposits will become scarce or expensive enough that we will be forced to use alternate energy sources. But this is at least several decades away and could take a century.

Because advances in extraction techniques tend to hold down future price increases, affordable fossil fuels will be available for many decades. The US has a 100 year supply of coal. Many countries have not yet begun to apply fracking technology to their gas reserves, suggesting similarly long-lived supplies of natural gas. Oil reserves beneath the deep ocean have just begun to be exploited, and within a few decades, the Arctic will be ice-free in summer, opening up that region to exploration. Canada's oil sands retain huge reserves. Well before scarcity curtails the use of fossil fuels, their environmental impact will force adoption of carbon taxes, emission controls, or other mechanisms to reduce their use and promote cleaner energy. The "great transition" to carbon-free energy will be painful for political, economic, and technical reasons, but it's a transition we must make if we care about our planet's ecosystems.

One of this chapter's themes has been the importance of understanding the relative risk of various energy sources, including long-term implications and full cost accounting of environmental impacts. During the great transition, we will have to balance the risk of economic disruption from using more costly alternative energy sources, against the risk of further environmental damage from fossil fuels. This is discussed in more detail in Chapters 8 and 9. Chapter 7 looks at flood hazards, and how one consequence of our overuse of fossil fuels (sea-level rise) is increasing this particular hazard along our coasts.

# 7 Past and Future Coastal Flooding: Galveston, New Orleans, Bangladesh, and the Specter of Sea-Level Rise

*"Fortune is an arbiter of half of our actions ... I liken her to one of these violent rivers which, when they become enraged, flood the plains, ruin the trees and the buildings ... It is not as if men, when times are quiet, could not provide for them with dikes and dams so that when they rise later ... their impetus is neither so wanton nor so damaging."*

Niccolo Machiavelli, *The Prince*, written in 1513, first published in 1532 (translation by H. C. Mansfield Jr)

Despite Machiavelli's prudent advice on flood mitigation (with some political advice thrown in for good measure), humans continue to be devastated by flooding. One of Britain's largest natural disasters in modern times, in terms of loss of life, was caused by a flood in 1953. A winter storm in the North Sea combined with high tides to raise local water levels by more than 5 meters along parts of the UK east coast, killing more than 300 people. Hurricane Katrina destroyed a significant fraction of the city of New Orleans in 2005, largely by flooding, generating damages in excess of $100 billion and crippling the presidency of George W. Bush. A similar storm (Hurricane Sandy) devastated coastal New York and New Jersey in 2012, with damages exceeding $50 billion. One of Canada's costliest natural disasters was associated with flooding in the province of Alberta from high rainfall in June 2013. Super Typhoon Haiyan, one of the strongest storms ever recorded, devastated the Philippines in November 2013. According to Dr. James Daniell, a researcher at Karlsruhe Institute of Technology in

Germany, floods have been the most damaging natural disaster of the 20th and early 21st centuries, accounting for roughly one-third of global losses.

In this chapter I'll discuss some aspects of the flood problem, using examples from Galveston and New Orleans in the US, and Bangladesh. I'll show why considering subsidence of the land surface can be just as important as sea-level rise when considering flood hazards.

Floods should be one of our more preventable disasters. Unlike earthquakes, where scientific understanding came relatively late, long after the establishment of most cities, we've known about flooding since antiquity. In fact, there's been published advice on how to avoid flood damage since 1532 (see Machiavelli's quote at the beginning of the chapter). Experts had been pointing out the dangers of flooding in New Orleans and New York many years before those events. Despite such foreknowledge, flood-related disasters hit the headlines almost every year. As shown in earlier chapters, they tend to be our most common and costliest disasters, and their costs are increasing with time. Clearly, we have not done a good job of following Machiavelli's advice.

Most flood disasters occur in river floodplains during spring melting or in coastal areas when tropical storms, hurricanes (big storms in the Atlantic basin), and typhoons (big storms in the Pacific and Indian Ocean basins) cause storm surges. While this chapter mainly focuses on coastal flooding, I will first make a few comments on overbank flooding adjacent to inland rivers. As noted in Chapter 1, this class of flooding should be avoidable – it always happens in the same place (the floodplain adjacent to the river) and mostly at the same time (spring melt season). Many countries worsen the problem by deforesting uphill slopes, which promotes rapid runoff of rain, or by allowing urban areas to expand onto their river floodplains. Urban construction styles also promote flooding from high rainfall events by reducing or eliminating green space that would otherwise allow rainwater to soak into the ground, where it can slowly percolate into the river. Instead of green space, we have covered our urban areas with

FIGURE 7.1 A large suburban parking lot near Pittsburgh, Pennsylvania, typical of many suburban areas in the US and other industrialized countries, with extensive hardscape. Elimination of green space reduces percolation of rainwater into the ground, forcing runoff to flow immediately into adjacent streams or rivers, causing flooding during high rainfall events. Note the oil patch in the foreground, a common surface deposit on city streets and parking lots. Runoff from these deposits includes toxic substances that course through our streams and rivers, eventually winding up in the ocean, and in the fish we eat. Photo by the author, taken August 2012.

asphalt, concrete, and other impermeable surfaces ("hardscape"; Figure 7.1). This forces water to immediately flow into adjacent streams or rivers, causing what I call "induced flooding," another one of those nasty "own goals."

In addition to relocating homes and sensitive infrastructure to higher ground, some communities are experimenting with more natural drainage systems to reduce induced flooding. For example, instead of parking lots with large, uninterrupted expanses of asphalt as in Figure 7.1, parking surfaces can be broken up by a series of green strips that promote percolation of storm water into the ground. These strips can be planted with trees, allowing some lucky car-parkers to find shade from hot summer sun. Storm sewers can also be designed to drain into adjacent wetlands rather than straight to the local stream or river. This has the added benefit of giving nature a chance to clean up the runoff before it enters the river, runoff that is often polluted with tar, oil, gasoline, and other substances that are an unfortunate by-product of our modern automobile-oriented society (Figure 7.1). My own city of St. Petersburg, Florida, has constructed a series of artificial wetlands. These have been quite successful, providing flood control and environmental clean-up services at relatively low cost, while also providing much-needed green space and wildlife habitat.

Hurricane-related storm surge can temporarily raise local sea level by 5–10 meters (16–33 feet) for several hours or even days, inundating coastal communities. Storm surge develops when strong winds blow toward the shore, and tends to be stronger in slower moving storms because the wind has more time to pile up water. Coastal flooding due to storm surge is usually accompanied by high winds, waves, and currents, and it devastates coastal structures unless they are deeply anchored. Most of the fatalities and destruction in the Philippines from Typhoon Haiyan in 2013 came from storm surge. That storm caused more than 6,000 fatalities and left more than a million people homeless. In 2008, Typhoon Nargis hit the low-lying Irrawaddy Delta of Myanmar (Burma), causing at least 100,000 fatalities in that county's worst natural disaster in recorded history. The actual number of fatalities is not well known. Many bodies were washed out to sea, and the country's military dictatorship limited access by reporters and foreign aid workers, suppressing news of the disaster and hampering relief efforts.

In this chapter I'll discuss two well-documented US examples (respectively, the disaster with the most fatalities and the costliest disaster) and then broaden the discussion. I'll look at the general problem of river deltas, where many of the world's great cities are located, and sea-level rise, an ongoing global phenomenon that will accelerate in the future and will increase the frequency of storm-related flooding.

## THE GREAT GALVESTON HURRICANE OF 1900

The town of Galveston, Texas, is located on a low-lying barrier island (essentially a large sand spit) in the Gulf of Mexico, south of Houston. On September 8, 1900, it was devastated by a hurricane estimated to be Category 4 on the Saffir-Simpson scale (Table 7.1). For comparison, Hurricane Katrina was probably a Category 3 when it struck New Orleans. Hurricane Andrew, which struck Miami in 1992, was a Category 5. Typhoon Haiyan in 2013 was also a Category 5 storm. Until recently, Haiyan held the record for the strongest sustained winds recorded for a storm as it approached landfall. Winds sustained over one minute reached 315 km per hr (195 miles per hr), estimated by the Joint Typhoon Warning Center in Pearl Harbor, Hawaii, using satellite data. The previous one-minute record had occurred during Hurricane Camille in 1969, with winds reaching 305 km per hr (190 miles per hr). In 2015, Hurricane Patricia off the coast of Mexico reached sustained winds of 325 km per hr (200 miles per hour), breaking Haiyan's record. Patricia also broke records for maximum wind speed for a hurricane (215 miles per hour) and minimum pressure

Table 7.1 *Saffir-Simpson hurricane scale*

| | Maximum Sustained Wind Speed | | |
|---|---|---|---|
| Category | Km/hr | Miles/hr | Damage |
| 1 | 119–153 | 74–95 | Minimal |
| 2 | 154–177 | 96–110 | Moderate |
| 3 | 178–209 | 111–130 | Extensive |
| 4 | 210–249 | 131–155 | Extreme |
| 5 | 250 or greater | 156 or greater | Catastrophic |

(872 millibars). There appears to be a pattern of increasing intensity as measured by maximum sustained winds, but El Niño (periods of warm water in the equatorial Pacific) also plays a role. It is also difficult to compare recent storms to ones like Galveston's 1900 event, since the earlier events were less well recorded.

At the time of the Galveston hurricane, the practice of naming major storms had not yet developed. The Galveston hurricane caused at least 6,000 fatalities. The official estimate is usually given as 8,000, between a quarter and a fifth of the town's residents at the time, making it the deadliest disaster in US history (Table 2.1). Even though Andrew, Haiyan, and Patricia were much larger storms, modern weather forecasting allowed many residents to flee dangerous coastal areas before these storms hit. In contrast, knowledge about the growth and frequency of hurricanes in 1900 was much less sophisticated than today; there was also limited ability to track storms and issue warnings. Nevertheless, a surprising amount of information is available on the storm; some tracking occurred, and some warnings were issued. It appears to have been a typical Atlantic hurricane, probably originating off the west coast of Africa, moving west across the equatorial Atlantic, gaining strength in the warm waters of the Caribbean and Gulf of Mexico, and after striking the Texas Gulf coast, moving inland to the north and east, eventually heading out to sea over Nova Scotia and Newfoundland in Canada.

Galveston's growth in the late 1800s accelerated after another coastal town, Indianola, was struck by hurricanes in 1875 and 1886. Some Galveston residents thought their town was similarly susceptible, while others argued that it was safe from hurricanes, as it had weathered a number of previous storms. A proposal to build a sea wall was defeated, in part because the director of the Galveston Weather Bureau, Isaak Cline, argued that it was unnecessary and that a major hurricane was unlikely to strike the island. When he wrote this in 1891, Galveston had been in existence for about 50 years and had never been struck by a major hurricane. Like many people before and since, Cline assumed that 50 years was a sufficient period of time to reliably indicate the future threat from natural disasters. The feeling of security from future storms extended to most of the town's inhabitants; sand dunes on the outer part of the island that afforded some protection from high water were mined for construction, effectively eliminating the town's only natural barrier to storm surge.

Surge from the storm that finally hit exceeded 4.6 meters (15 feet), well above the highest point of land on the island, 2.7 meters (8.8 feet). Buildings were pushed off their foundations by the surge and then broken apart by the winds and waves. Only a few solidly built and anchored structures survived. Many bodies were buried under the rubble of their houses. After the storm, recovered bodies were burned on the beach for many weeks.

One could argue that barrier islands are not ideal places to build towns and cities, because of their low elevation and exposure to tropical storms and hurricanes. Perhaps they are better left as parks and green space for the enjoyment of everyone and for environmental refugia. But if barrier islands are going to host urban populations, infrastructure needs to be engineered correctly. Several years after the storm in Galveston, a sea wall was finally constructed. In addition, the town's base level and most structures were elevated using dredged sand. These measures were costly (the population, with the recent disaster in mind, was willing to make the investment) and ultimately effective. A second hurricane struck the town in 1915, but this one caused much less damage and very little loss of life. Galveston was not devastated again by a hurricane until Hurricane Ike in 2008. That storm caused billion of dollars of damage but few fatalities.

The 1900 Galveston hurricane played an important role in our understanding of hurricanes and it motivated improvements in the US Weather Service, hurricane tracking, mitigation strategies, and the general level of the nation's preparedness for hurricanes. While the southern parts of the US would continue to be hit by hurricanes, never again would one cause so much devastation, or catch the US by surprise – until Katrina.

## NEW ORLEANS AND HURRICANE KATRINA

New Orleans is located on part of the low-lying Mississippi River delta. It is unique among US coastal cities in having a significant fraction of its land area below sea level, which makes it especially susceptible to flooding. On August 29, 2005, the city was struck by Hurricane Katrina. Although Katrina had earlier intensified to a Category 5 hurricane as it passed over the warm waters of the Gulf of Mexico, its strength had probably declined to a Category 3 by the time it struck the city.

While most residents were warned well in advance of the storm's landfall, some were unable or unwilling to evacuate. Of the approximately 1,800 fatalities associated with Katrina, approximately 1,000 were residents of New Orleans who had remained behind. Many deaths were attributed to drowning. Katrina's main impact on the city was the breeching of the city's levees, which previously had protected the city from flooding by the Mississippi River and Lake Pontchartrain. The city remained flooded for many weeks, slowing rescue, relief, and recovery efforts. This hurricane is generally regarded as the costliest disaster in US history and the second most deadly hurricane after Galveston. As of 2014, the population of New Orleans was about 75 percent of the pre-storm level. Parts of the city most devastated by the storm remain in a damaged state. Before discussing how such a disaster could afflict a wealthy and technologically advanced nation like the US in modern times, we need to discuss some geologic background.

## RIVER DELTAS, SUBSIDENCE, AND LEVEES

Deltas are basically big piles of sand, silt, and mud that are deposited where a river meets the ocean. The river currents slow down, and sediments carried by the river settle out. When viewed from above (Figure 7.2) many deltas look a bit like a triangle, or the Greek letter delta (Δ is Greek for "D"). The surfaces of coastal deltas lie very close to sea level. As new sediment comes in (e.g. brought in by

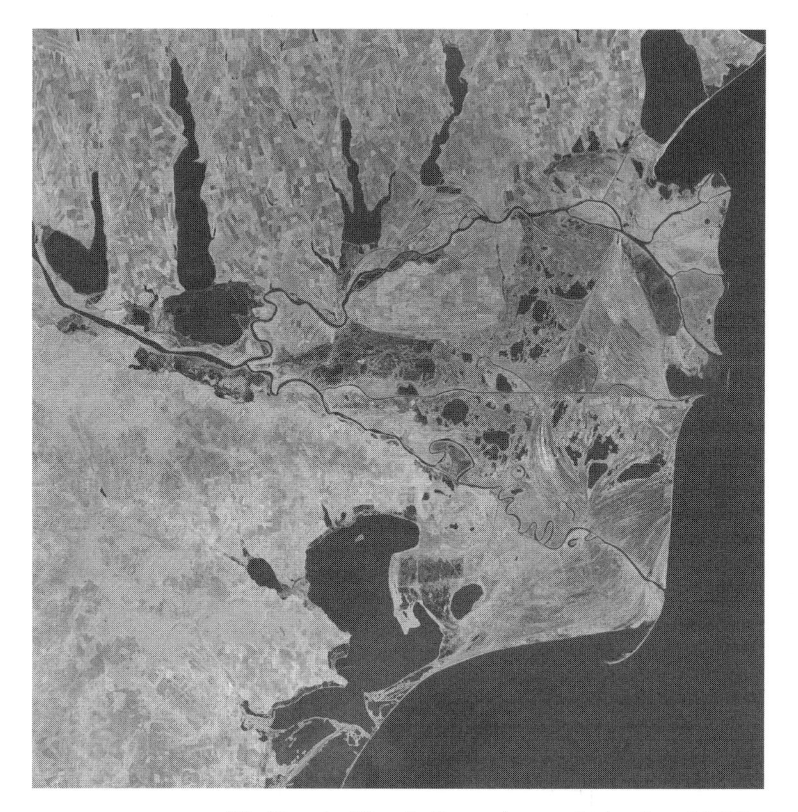

FIGURE 7.2 The Danube River in Romania, near its border with Ukraine, ends in a classic delta shape, where the river empties into the Black Sea. It is one of the largest deltas in Europe, and parts of it have been designated a World Heritage site due to unique wetland ecosystems. This image is from NASA's Landsat satellite. Image courtesy of NASA.

spring floods), it gets deposited on top of older sediment, eventually compressing the underlying layers of mud and silt. However, as the surface subsides due to this compaction, the next batch of sediment comes in. A dynamic process is established with ongoing compaction compensated by ongoing sedimentation, constantly re-surfacing the delta to maintain its elevation close to sea level. The process can go on for many thousands of years, eventually building a huge delta that sticks out from the surrounding coastline.

Deltas were considered ideal places to build towns in the past, before we knew much about their longer-term geology. They represent a meeting place for river transport, bringing goods to be exported from the interior of a continent, and ocean transport, bringing goods to be imported from overseas countries. Before the advent of airplanes, deltas were the world's transportation hubs. It's not surprising that some of the world's great cities, such as Alexandria, London, New York, and Shanghai, are built on or near river deltas.

New Orleans lies on a delta and parts of it are very low lying, even below sea level, in some places more than 3 meters (10 feet) below sea level. This is related to several natural processes that result in surface subsidence, combined with the long-term effects of later levee construction and sea-level rise. The levees eliminate spring flooding, but have the unfortunate side effect of reducing sedimentation that would normally come with the floods. Without the levees, spring flooding would deposit new sediments on the top of the delta, maintaining the surface close to sea level. Instead, the channelized Mississippi River now dumps its sediment far out to sea. The subsidence continues, but not the compensating sediment deposition, so the surface gets lower all the time.

We can actually measure this process with modern satellite techniques – some parts of the city are subsiding at rates of 20–25 mm (about 1 inch) per year (Dixon et al., 2006). A simple calculation shows that the current low elevations in the city are related to the rate of subsidence, assuming it began shortly after levee construction. The lowest elevations in New Orleans today are about 3 meters (10 feet) below sea level, The earliest levees were constructed in the mid 1800s, so let's assume that some surfaces started subsiding from near zero elevation (close to sea level) in 1860. In 2005, the year of Katrina, that means 145 years worth of subsidence. Over the course of 145 years, 3.0 meters of subsidence gives an average subsidence rate of about 21 mm/yr (nearly one inch per year), which is close to the maximum rate measured by satellites.

Like many things in science (and life) the story is more complicated. Other factors also contribute to subsidence and low elevation in New Orleans and the Mississippi Delta, including past extraction of oil and natural gas, pumping of groundwater, and regional subsidence caused by the load of the Mississippi Delta on the Earth's crust and lithosphere. To understand today's low elevations, we need to consider all the various processes that cause subsidence, consider how long they have operated, and add them up. The highest subsidence rates measured today occur in areas of artificial fill along the lakeshore and wetland areas that were drained for urbanization in the 1920s and 1930s. Many of the lowest areas actually have relatively low subsidence rates today (they are still subsiding, but at lower rates). This reflects the fact that subsidence is also caused in part by compaction and oxidation of young, organic rich wetland and marsh deposits, and these processes slow with time. The simple calculation in the previous paragraph uses an average subsidence rate and is only a first approximation to a complex problem.

Given the standard saying among engineers about the two types of levees (those that have failed, and those that will fail), many geologists think it would be prudent to move infrastructure in the lowest-lying parts of New Orleans to higher ground. These areas remain susceptible to catastrophic flooding during tropical storms and hurricanes, when storm surge can overtop the levees and flood the city. The lowest-lying parts of the city (lower than 2.5 meters below sea level – they can even flood in a heavy rain) represent a relatively small percentage of the broader urban region. Many parts of the city are actually at or above sea level. Rebuilding infrastructure that currently lies at the lowest elevations on higher ground should be feasible. The lowest-lying areas are of course most vulnerable, and when flooding occurs, it can be fatal for people who can't swim and are not able to reach higher ground or a tall structure. Many fatalities in Hurricane Katrina involved people trapped in the attics of single-story houses as they struggled to escape rising water.

## COULD IT HAPPEN AGAIN?

Unlike Galveston after the 1900 hurricane, New Orleans is not able to raise its base level above current sea level. The city is simply too large. On the other hand, it is possible to abandon the lowest areas, focus rebuilding on higher ground, and require new construction to be flood-resistant, for example through the use of an open design for the ground floor (e.g. for storage and parking), with the first residential floor elevated to the second story. This design is common in many flood-prone areas such as the Florida Keys and Mississippi Delta (Figure 7.3).

After Katrina, it would have made sense for the state and federal governments to buy out homeowners in the lowest-lying areas, or subsidize reconstruction that employed flood-resistant designs. Such a rebuilding and relocation effort would have represented a political

FIGURE 7.3 A typical Delta house with the main living space on the second floor. Photo by the author, taken March 2016.

challenge, but it might have been possible in the presence of a strong scientific consensus and a clear, simple statement of the problem that could be understood by the general public. In the aftermath of that disaster, New Orleans was devastated, and the population was in shock, understandably suspicious of a government that had fumbled the relief effort. Some clear, common-sense advice from an independent scientific body would have been useful.

The US Geological Survey, the American Geophysical Union, and other scientific groups were asked to give advice on the scientific background to the catastrophe. A study was initiated involving a large group of experts, with specialists on hurricanes, weather forecasting, climate change, geologists (including myself), engineers, and policy experts. The report (*Hurricanes and the U.S. Gulf Coast: Science and Sustainable Rebuilding*) was published in 2006 and was a complete and authoritative analysis of the scientific background to the disaster. Here are the first two sentences of the paragraph in the Executive Summary concerning elevations, subsidence, and future flood susceptibility: "Natural processes as well as human impacts have contributed to subsidence, the sinking of land over time, along the Gulf Coast. Presently, there is considerable discussion and debate among the scientific community regarding mechanisms and rates of subsidence in the Mississippi delta area."

The body of the document goes on to accurately define the problem, but for policy makers with short attention spans – who may only read the Executive Summaries of such reports and might only read those first two sentences – the conclusion would probably be "nothing useful here, the usual scientific waffle; let's ignore it." What could have been a clear statement on low elevation and future flood susceptibility in parts of New Orleans instead focused on discussions about the various natural and human-induced processes that led to low elevations, the rates of subsidence, and the uncertainty of some of the data, none of which are directly relevant to the flooding problem. As with other committee statements highlighted in this book, clarity was sacrificed for consensus.

Given the lack of clarity from scientists on this issue, it is perhaps not surprising that there was little political enthusiasm for relocating sensitive infrastructure to higher ground or requiring flood-resistant construction such as elevated structures. While some of this is happening, many residences in New Orleans were rebuilt to essentially their pre-storm elevation. By sensitive infrastructure, I mostly mean housing for the poor, who tend disproportionately to live in marginal, low-lying areas prone to flooding.

I still feel guilty about my inability to sway my peers or influence the tone, writing style, and substance of the 2006 report. To make up for it, here is my preferred summary concerning elevations, subsidence, and future flood susceptibility in New Orleans, without the inevitable compromises imposed by committees:

> More hurricanes, and more intense hurricanes, are coming, and the levees will fail again at some point in the future. Single-family residences in low-lying areas need to be elevated for flood-resistance or moved to higher ground. The lowest-lying areas should be allowed to revert to green space. State and federal assistance should be targeted at those most in need to facilitate this difficult transition.

Many people believe that it is unrealistic to allow the lowest-lying areas of New Orleans to revert to green space. People would refuse to move (geographic inertia), and in any case, it would be too expensive to subsidize the transition (financial inertia). The first issue is certainly challenging, but it would be interesting to see how people responded to the following offer: Everyone gets a heavily subsidized opportunity to move to a safer area. Vacated areas become green space, and former residents retain ownership of this green space, which could become individual or community gardens, a dedicated park, or a community center. The second issue (cost) is actually much easier. Let's assume there are approximately one million people in the greater New Orleans area, and 20 percent (200,000) live at dangerously low elevation. Lets further assume that this represents 50,000 families (i.e. the average family consists of four people) and that houses or condominiums for this hypothetical family of four could be purchased elsewhere in the city for $150,000, a typical value for the area. Assuming the federal government paid for everything, that works out to a total of $7.5 billion. For comparison, the US government spent hundred of billions of dollars during the 2008–2009 recession, bailing out banks, other financial firms, and companies like General Motors (GM), despite their history of poor fiscal management and their overpaid executives. For example, GM received approximately $50 billion. Most of this was eventually recouped by the government, but taxpayers wound up losing an estimated $11 billion on the GM deal, more than the cost of buying out everyone living at low elevation in New Orleans.

This brings up an important societal question, one that will become increasingly important as governments decide how to handle future "retreats from the coast" that in the long run are an inevitable consequence of sea-level rise. When government assistance, subsidies, or "bailouts" are being considered, it is often difficult to strike the right balance. In the case of financial crises, a mainstay of Keynesian economics is that it is important for governments to intervene and keep banks and other financial firms solvent, maintaining credit flows and preventing an economic "death spiral." On the other hand, it is also important not to reward bad behavior. Economists use the term "moral hazard" to describe the obvious problem of governments appearing to bail out rich executives and the companies they mismanaged, misbehavior that likely contributed to the financial crisis in the first place. To minimize moral hazard, strings are usually attached to financial aid. For example, top executives may lose their jobs, and creditors and shareholders have to take a "haircut" (a loss). A similar balance has to apply to bailing out owners of flood-prone properties. On one hand, it seems to me that low income residents of low-lying areas in places like New Orleans deserve a subsidized transition to high ground if they choose to move. On the other hand, how should governments design subsidized relocation programs so that they don't wind up benefiting wealthy owners of pricey oceanfront real estate?

Some of the problems at the Fukushima Daiichi nuclear complex have striking similarities to earlier problems experienced by some New Orleans hospitals and other facilities after Hurricane Katrina. In the latter case, backup diesel generators or critical control systems were located on ground floors or even in basements, and were flooded when storm surge over-topped levees. The designers of these backup power systems apparently could not conceive of the possibility of a flood, even though geologists had been warning of the likelihood of flooding in New Orleans for many years. In fact, parts of New Orleans were flooded by Hurricanes Betsy and Camille in 1965 and 1969, respectively, in eerily similar events. The possibility of flooding should have been in everyone's mind, certainly in the minds of architects, designers, and builders of emergency backup systems in hospitals. Unfortunately, lessons from several decades earlier were forgotten.

Other coastal areas also experience subsidence and vulnerability to flooding. Jakarta, the capital of Indonesia, experiences frequent flooding and has some of the same problems as New Orleans: a low-lying delta location (where the Ciliwung River enters Jakarta Bay) and a high subsidence rate, with many parts of the city getting lower by several centimeters (about 1 inch) per year. Unlike New Orleans, the hills behind Jakarta and the area's natural drainage are also quite effective at channeling high seasonal rainfall events directly into the city.

The subsidence experienced by the world's coastal cities reflects several geologic processes, including oil and gas withdrawal (a problem along parts of the Gulf of Mexico coast), excess withdrawal of groundwater (a problem in many urban areas), compaction of artificial fill (coastal urban areas often try to expand their oceanfront real estate using dredged material), and oxidation of organic-rich marsh and delta deposits (a problem in both New Orleans and Jakarta). While the rates of subsidence are usually low compared to New Orleans and Jakarta, in the long term, all of these areas will experience similar hazards from storm surge through a combination of subsidence and sea-level rise.

Nuclear plants and other critical facilities located in coastal areas need to include the combined effects of subsidence and sea-level rise in their planning.

## SEA-LEVEL RISE: A SLOW DISASTER IN THE MAKING

Sea-level rise is the inevitable response to global warming, which is discussed in the next chapter. During most of the 20th Century, sea level rose at a global average rate of about 1.5 to 2 mm/yr (e.g. Figure 2.7). At the present time, it appears to be rising faster. Satellite measurements suggest a current global average rate of more than 3 mm/yr (Figure 7.4). Locally, some areas experience much higher rates of sea-level rise.

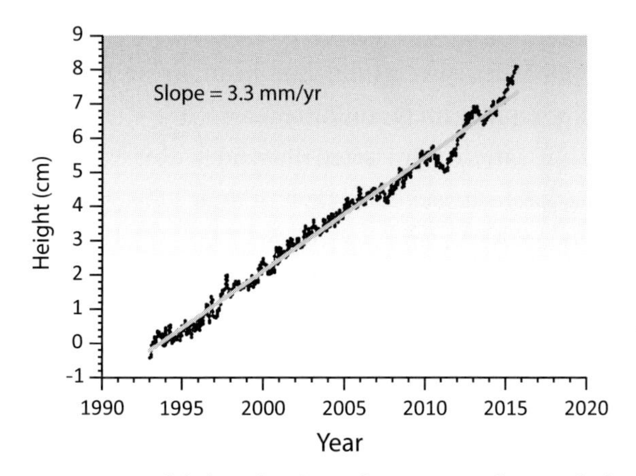


FIGURE 7.4 Global sea level was first measured precisely from satellite in 1979 by NASA's Seasat satellite. Since 1992, it has been measured more or less continuously by a suite of satellites from various nations. The technique involves a radar altimeter to measure the height of the satellite above the ocean surface, and precise orbital tracking to correct for satellite location and motion. Unlike tide gauges which are restricted to a relatively small number of coastal locations, satellites obtain full global coverage. The satellite data show a higher rate of sea-level rise (approximately 3.3 mm/yr) compared with the 20th Century average (1.5–2.0 mm/yr, e.g. Figure 2.7), suggesting that sea-level rise is accelerating. Data courtesy of the French space agency CNES (Centre National d'études Spatiales).

Sea level rises in response to a warmer atmosphere for three main reasons. First, a warming atmosphere warms the ocean. Most things, including the ocean, expand in volume when they warm up, a process called thermal expansion. Plumbers who have to fit two pipes together will often heat the wider pipe with a blow torch. The heated pipe temporarily increases in size and slides more easily over the narrower one. Second, alpine glaciers (the small ones near mountain tops) start to melt faster, adding water to the ocean basins. Third, the great polar ice sheets in Greenland and Antarctica also begin to melt. For most of the 20th century, sea-level rise was mainly due to the first two processes. In fact, many scientists assumed that Greenland and Antarctica were unlikely to melt any time soon, regardless of the amount of global warming, because their great mass gave them a sort of thermal inertia, and they were located in cold, polar regions. Beginning in the 1990s however, many outlet glaciers on the edge of the great ice sheets that had been stable for decades or centuries began to rapidly melt and retreat.

The GRACE satellite mission illustrates this recent change. GRACE is remarkable in several ways. It is a joint space mission between the US (NASA) and Germany (DLR, Germany's space agency). GRACE was launched in 2002 from Russia's Plesetsk Cosmodrome facility in Siberia, a facility originally developed by the Soviet Union during the cold war to launch intercontinental ballistic missiles in the event of a hot war. The GRACE mission consists of two satellites, GRACE-A and GRACE-B (sometimes called Tom and Jerry) that orbit at around 500 km altitude (Figure 7.5). The two satellites are in essentially the same orbit, but they are separated by about 200 km. The satellites are tracked with onboard GPS units that allow the positions of the two satellites to be known to better than a centimeter. What is even more remarkable is that the distance between the two satellites is measured by a microwave ranging system (in some ways like a radar) to extremely high precision – several micrometers (a micrometer is one millionth of a meter). This precision in the distance measurement is what enables GRACE to do its job, which is to

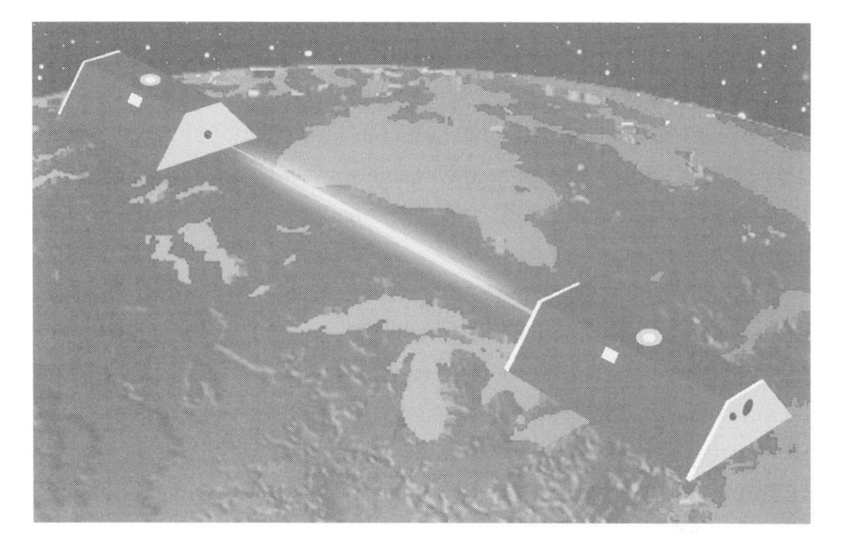

FIGURE 7.5 Artist's conception of the twin GRACE spacecraft in orbit. The two spacecraft are separated by about 200 km (125 miles) and a sensitive microwave system (basically a radar) continuously measures the distance between them. Image courtesy of NASA.

measure the Earth's gravitational attraction and, more importantly, *changes* in gravity.

To understand how this works, think about two free-wheeling, unconnected cars on a roller coaster. As the front car approaches the top of a hill, it slows down, and the rear car gets a little closer. As the front car rolls over the top of the hill, it speeds up, moving farther away from the rear car until the rear car also goes over the top of the hill and catches up. The distance between the cars is always changing, reflecting the relative gravitational accelerations each one feels as a result of the hills and valleys of the roller coaster track. In the same way, the twin GRACE spacecraft get closer or farther apart depending on the distribution of mass beneath them on the Earth's surface and in its sub-surface. The mass distribution causes changes in local gravitational acceleration. The twin GRACE spacecraft feel these changes, even at an altitude of 500 km, and speed up or slow down accordingly, changing their relative distance, albeit by tiny amounts. Measuring these changes in distance to an accuracy of a few micrometers (much smaller than the width of a human hair) allows scientists to determine the gravitational field beneath the spacecraft.

GRACE makes a map of the Earth's gravitational field roughly once month, giving a pretty good idea of how Earth's mass distribution is changing. While the rocks don't change much in a month (except during big earthquakes such as the ones that struck Sumatra in 2004 and Japan in 2011), Earth's fluid envelope does. Groundwater can change a lot (think of India's seasonal monsoon, or wet and dry seasons in the Amazon basin). More importantly for this chapter, the mass of Greenland and Antarctica changes due to melting of their respective ice sheets. They change on an annual cycle, reflecting summer loss and winter growth of ice, and on longer time scales as well (Figure 7.6). The biggest change observed by GRACE so far is Greenland. Greenland's ice sheet is melting away much faster than experts had predicted even 15 years ago, so fast that it will likely be the main contributor to global sea-level rise within a decade or two.

## SLOW SEA-LEVEL RISE AND RAPID FLOODING

What will the effects of sea-level rise look like in the next few decades? Probably much like the last few decades. Individual flood events will strike several low-lying places, with potentially devastating consequences. In 1991, a massive cyclone struck Bangladesh. Much of this country's population lives on a low-elevation river delta, where the Ganges and Brahmaputra Rivers empty into the Bay of Bengal. As with the Mississippi Delta, it is slowly subsiding. Winds from the cyclone drove storm surge far inland. The combination of high winds, flooding, and the subsequent disease and starvation led to approximately 130,000 fatalities and devastating economic losses for this already impoverished nation. Massive flooding again devastated the country in 1998, this time associated with high rainfall upstream. Deforestation in the upstream areas also contributed to flooding. Densely forested hill slopes reduce rapid runoff in high rainfall events, encouraging absorption of rain into the soil. Cutting down the trees

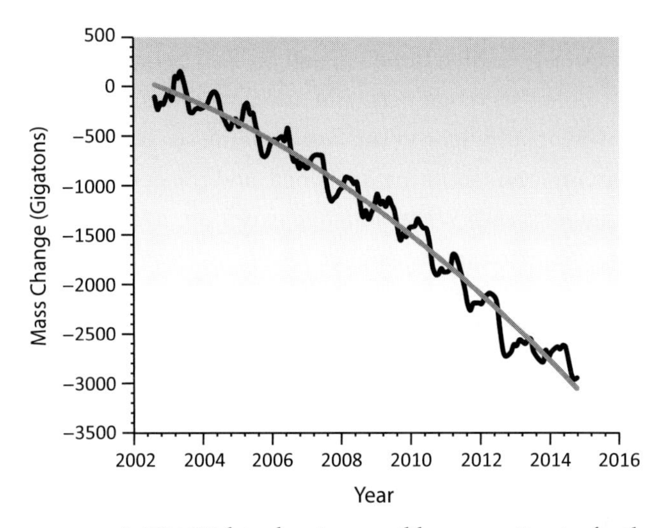


FIGURE 7.6 GRACE data showing monthly mass estimates for the Greenland ice sheet (small black circles) between 2002 and 2014, arbitrarily setting the starting point (the average of the first year's measurements) to zero. Note the annual changes, reflecting summer melting and winter growth, and the longer term loss of ice (the downward trend of the graph). Note also that the losses are accelerating (the graph curves downward). During this particular period, Greenland lost more than 2,500 gigatons (2500 billion tons) of ice, roughly 0.1 percent of its total. During the first five years of the mission, Greenland lost mass at an average rate of about 200 gigatons (GT) per year. During the next five years, the average loss rate increased to about 300 gigatons per year. The thin line shows a model fit to the data assuming a constant annual cycle (winter growth, summer loss) and constantly accelerating long term mass loss at 20 GT per year per year; i.e. every year, the rate of loss increases by 20 GT/yr. Modified from Yang et al. (2016).

allows rainwater to flow rapidly downhill, worsening flooding downstream. As sea level continues to rise and storm intensity increases (see Chapter 8), flooding events in Bangladesh similar to the 1991 and 1998 catastrophes are virtually certain to increase in frequency and severity.

The 2005 flooding of New Orleans in the aftermath of Hurricane Katrina, and the 2012 flooding of New York and New Jersey associated with tropical storm Sandy, give clues to what the future will look like for coastal parts of the US. Many levees in New Orleans were rebuilt in the 1960s after extensive flooding, and for decades, they seemed to be working. Geologists, engineers, and urban planners warned that such defenses were insufficient in the face of rising sea level and subsiding land. For many years, nothing happened, and the optimists seemed to have the upper hand. When Katrina struck in 2005, the associated storm surge, starting from a higher base level (several decades worth of sea-level rise) attacked levees that were now too low (several decades worth of subsidence). In the end, coastal defenses were overwhelmed from over-topping and other failure modes, including erosion at the base of levees. Experts had warned of New York's and coastal New Jersey's vulnerability to storm surge for a long time before Hurricane Sandy struck in late October 2012. As with Bangladesh, these events will hit the US more frequently in the future as sea level slowly rises and storms become more intense. Areas that also experience coastal subsidence, such as cities built on river deltas or coastal areas built on dredged fill, are more likely to be hit first and suffer more extensive damage when storms do hit.

Most people (including many scientists) assume that this is all that most coastal communities will have to deal with for the next 50 years or so in terms of hazards and costs related to sea-level rise – the occasional violent storm and associated flooding. Catastrophic, to be sure, but rare enough that individual communities can recover and rebuild. Rebuilding costs will be steep, but bearable: more than $100 billion for New Orleans and more than $50 billion for New York and New Jersey. Even the most pessimistic estimates predict that sea level will rise by less than 0.2 meters (a little less than 1 foot) by 2050, hence the amount of rise and additional risk is small compared to the short term but much larger effects from 5–6 meter storm surge. It's the rapid storm surge, not the slow sea-level rise, that's important, at least in the short term.

However, there are also more subtle hidden costs associated with sea-level rise. In their book *The Battle for North Carolina's Coast*, Dr. Stanley Riggs, a professor at East Carolina University, and several colleagues investigated the effects of sea-level rise on the health of their state's beaches, a major tourist attraction and recreation resource for the public. In many places, private housing was built behind the public beaches decades ago. However, rising seas are eroding the beaches, bringing new coastline to the edge of private developments, often protected behind sea walls and sandbags. The net effect is that the public beaches are increasingly lost. Long-term loss of tourist revenue is a likely outcome.

Miami Beach, Florida, a beautiful, art-deco tourist destination and a major source of income for South Florida, illustrates another example of hidden costs. When I was a professor at the University of Miami, a graduate student I knew lived in Miami Beach from 2005 to 2008 and parked her car on the street near her apartment. On several occasions, street-level flooding was bad enough to rise above the door panels of the car and flood the interior, requiring professional cleaning. On one occasion, flooding was severe enough that the entire engine had to be replaced. The car's electrical system also became heavily corroded, requiring additional repairs. While insurance covered most of her losses, over a four-year period she experienced several thousand dollars in unreimbursed costs from deductibles, co-payments, and lost time.

Parts of Miami Beach and Fort Lauderdale now experience flooding several times a year. These events used to be restricted to periods of intense rainstorms. The storm sewer system does not have much gradient, so during intense rain, the water has no place to go and fills the streets. In the last decade, a new phenomenon called sunny-day flooding has started to occur (Figure 7.7). If high tide corresponds with a period of offshore wind, local sea level rises enough to flood some of the streets, even without rain. Saltwater comes up through the storm sewers, and brackish water (mixed freshwater and saltwater) wells up through the porous ground.

If you were living in Miami Beach and wanted a safe place to park your car, you might consult a topographic map (a map with elevation information), or if you have access to a computer or

FIGURE 7.7 **(top and bottom):** Two pictures of Miami Beach during the "king tide" of October 2014. In the bottom picture, note that sea level is already quite close to the ground level – even a moderate storm or onshore wind would lead to significant inundation. Courtesy of S. Wdowinski (top) and Q. Yang (bottom).

smartphone, you could consult Google Earth. Google has digitized the topographic maps of many areas, so it's possible to read off the elevation of a specific location. If you are using an older map, the elevations will be in feet, and many areas of Miami Beach will lie on or close to the three-foot level (slightly lower than one meter). On Google Earth, they show up at or near the 1 m elevation mark. The elevation measurements are relative to mean sea level, defined as the average sea level over several decades, using a ground reference or "datum" called NAVD-29 (North American Vertical Datum 1929). You might think that as long as you were lucky enough to find a parking space for your car that was higher than 3 feet in elevation, which means three feet above the average height of the ocean, you would be safe most of the time, except during hurricanes or other extreme events. But, you would be wrong for several reasons.

First, the datum is now incorrect. Topographic maps are constructed with the assumption that the Earth is static – neither the Earth's surface nor sea levels are supposed to change. For most places most of the time, static Earth is a good assumption. Most people (except geologists) think of the Earth in this way. But the Earth is actually changing all the time. In the more than 80 years since NAVD-29 was defined and heights in Miami Beach were measured relative to it, certain parts of Miami Beach (the ones built on artificial fill) have subsided, reflecting compaction of the fill (by one or two feet), while sea level has risen by at least a foot. So places originally deemed to be 3 feet above mean high tide are now much closer to average sea level (i.e. zero elevation). If the tide is higher than average, the streets will flood. Mean sea level is updated every few decades, but maps tend to be updated more slowly. Many buildings in Miami Beach, including most of its art-deco hotels, were built in the 1920s and 1930s, using maps based on NAVD-29 or earlier datums.

Second, as the surface of a street gets closer and closer to sea level (approaching zero elevation) details become important in terms of flood potential. Small height differences (a foot or two) can make all the difference. Local depressions or high points may not be recorded on a typical topographic map. Also it is no longer enough to know the average elevation. This close to zero elevation, we also need to know the detailed time variation of sea level, which can change several tens of centimeters (1 to 2 feet) in an hour or less, due to tides and local weather. Miami Beach does not have levees like New Orleans to hold back high water, so high water conditions are felt almost immediately, with water often forcing its way up through the storm sewers or porous ground. High tide actually varies quite a bit. For Miami Beach, a spring high tide (when the moon is full or new, aligning its gravitational pull with the sun) will be a foot or two higher compared to a neap high tide (when the sun and moon are at right angles). The difference between spring and neap tides does not matter if you live in Colorado, much higher than sea level, but if you are living close to the ocean, that extra foot or two can make the difference between flooding and no flooding. Sometimes your parking space is actually going to be one or two feet below that day's average water level, and you need to take that into account if you are looking for a safe place to park. In effect, you need a weather forecast for local sea level. If you knew sea level was going to be especially high over the next 24–48 hours, due to either a spring tide or an offshore wind, and you had the option, you might choose to pay extra to park your car overnight on the second or third floor of a parking structure.

The third reason topographic maps are misleading in terms of flood potential is related to the second. As the street surface gets lower and lower, it no longer drains very well. A heavy rainstorm can lead to significant local, short-term flooding until excess water has time to drain away. If the rain happens to coincide with high tide, things get much worse since there is nowhere for the water to drain until the tide recedes.

Perhaps this will be the next new thing for software developers – an app for coastal residents that details the local high spots, and predicts actual sea level and flood susceptibility for various locations for a given time of day, similar to a weather forecast. The prediction would be based on local elevation and drainage, short-term changes

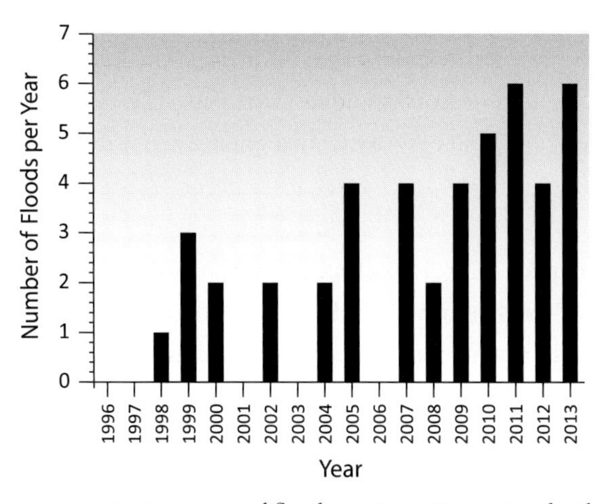


FIGURE 7.8 Frequency of flood events in Miami Beach, Florida. Modified from Figure 3a of Wdowinski et al. (2016).

from tides and weather, and longer-term changes from sea-level rise and land subsidence. Available data suggest a significant increase in the rate of sea-level rise in the last decade, but it is still not completely understood. Dr. Shimon Wdowinski, a professor at Florida International University in Miami, has studied the frequency of local flooding in Miami Beach using media reports, insurance claims, and weather records (Figure 7.8). There has been a rapid increase in the number flood events over the last two decades.

Frequent flooding of Miami Beach with water that is increasingly saline does more than damage parked cars. It is starting to wreak havoc with all sorts of infrastructure, from building foundations to buried cables and underground pipes for water and sewage. Gravity-drained storm sewers have become ineffective and require costly pumps, similar to those used in New Orleans. The present and future cost of these repairs and infrastructure upgrades is very high. In 2013 and 2014, Miami Beach spent million of dollars installing new pumps to flush floodwaters into nearby Biscayne Bay. Unfortunately, this "fix" has had unintended economic consequences. Biscayne Bay is an important tourist attraction, famous for its swimming, boating, and fishing. The untreated floodwater is hundreds of times higher in *enterococci* (the kind of bacteria that indicates fecal contamination) compared to levels recommended by the US Environmental Protection Agency. This is not exactly a great tourist draw.

## WHAT OTHER REGIONS ARE AT RISK?

We can also expect more flooding along parts of the US Eastern Seaboard, from Washington DC to the Carolinas, due in part to an additional process: glacial isostatic adjustment (GIA) (Figure 7.9). This natural process is a hangover from the northern hemisphere's last major glaciation, which ended roughly 8,000 years ago. Prior to that time, thick ice sheets covered the northern half of the continent. At its maximum, around 20,000 years ago, the ice sheet depressed the land under the glaciers by 100 meters (300 feet) or more due to its great

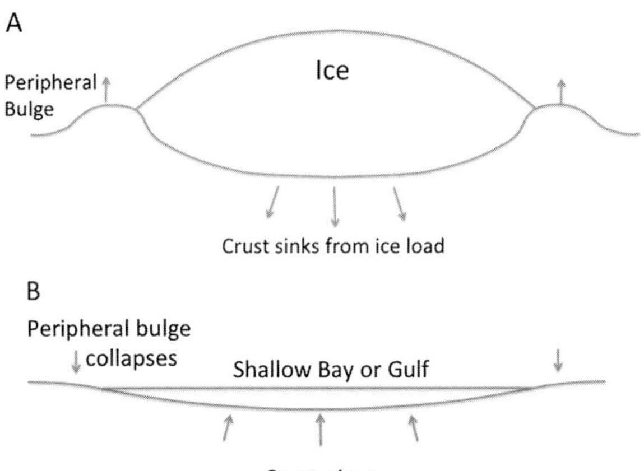


FIGURE 7.9 Glacial isostatic adjustment (GIA). **A (top):** Glaciers load the northern part of a northern hemisphere continent (North America or Scandinavia), reaching a maximum about 20,000 years ago. Their weight depresses the landscape beneath, but peripheral areas are slightly elevated. **B (base):** After the glaciers melt, the land beneath them begins to rise; the residual depression becomes a shallow bay (e.g. Hudson Bay in North America). The peripheral bulges slowly collapse. Both the central uplift and the peripheral bulge collapse continue today.

weight. However, land on the edges of the ice sheet was actually uplifted, including much of the Eastern Seaboard south of New York. As the glaciers receded, the weight was removed, and the land underneath the glacier rebounded, slowly returning to its original elevation while peripheral areas (those located on the "peripheral bulge") collapsed. Because the underlying layer of the Earth, the mantle, is quite viscous (think of a fluid like cold molasses) it flows slowly, so the adjustment is not instantaneous. In fact it's still going on. Land around Hudson Bay is currently rising at about 10 mm/yr, while parts of the eastern seaboard of North America are subsiding at rates up to 2 mm/yr. That's close to the rate of current sea-level rise (3 mm/yr) so the net effect is that the flood potential for parts of the eastern seaboard is increasing at rates of up to 5 mm/yr (3 mm/yr of sea level going up, plus 2 mm/yr of land going down).

The city of Norfolk, Virginia, experienced little or no flooding until the 1980s. In 2003, parts of that city experienced serious flooding associated with strong offshore winds. Additional floods occurred in 2009 and again in 2012, the latter associated with Hurricane Sandy. It is unlikely to be the last flood – Norfolk has four strikes against it. In addition to low elevations characteristic of this coastal region and the global sea-level rise that all coastal communities experience, parts of Norfolk are also built on compacting delta sediments. Finally, Norfolk happens to lie near the bull's eye for peripheral bulge collapse (Figure 7.9b) on the Eastern Seaboard. The net result is that parts of Norfolk experience relative rates of sea-level rise (the sum of all the land subsidence processes plus local sea-level rise) of 5–7 mm/yr or more. By the 1990s, enough subsidence had occurred that buildings constructed in the previous century were now at dangerously low elevations. Low-lying homes and businesses near the coast can expect to experience heavy economic losses in the decades ahead, from individual flood events and from the ongoing cost of flood insurance.

A similar process of peripheral bulge collapse affects parts of eastern Great Britain and the North Sea coasts of Germany and the Netherlands. Nocquet et al. (2005) estimate subsidence rates of 1–1.5 mm/yr from this process. While this seems quite slow, recall that towns and villages were established much earlier in Europe than in North America. A coastal European site that may have been ideal for settlement a thousand years ago would today be roughly a meter lower in elevation in areas affected by peripheral bulge collapse. This may have been a contributing factor in Great Britain's 1953 flood disaster, which was mentioned at the beginning of the chapter.

An early draft of this chapter, written in 2011, stated that in the lifetime of our children or grandchildren, New York would likely experience severe flooding from a tropical storm, similar to what happened to New Orleans during Katrina. Hurricane Sandy in late October 2012 moved my timeline forward. That storm caused more than 100 fatalities and at least $50 billion in damages to New Jersey and New York.

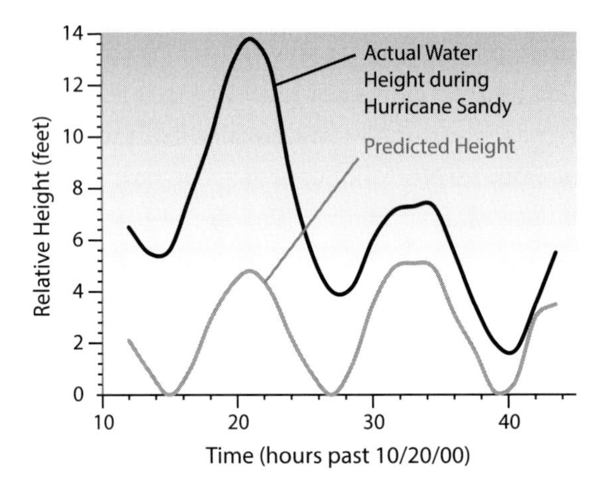


FIGURE 7.10 Water level at The Battery, New York, as measured by a tide gauge, showing storm surge associated with Hurricane Sandy beginning at midnight (00:00 hours) on October 29, 2012. At the height of the storm, local sea level was 9 feet (nearly 3 meters) above normal. Note that water levels stayed elevated for more than a day. Data courtesy of NOAA/NOS.

The Saffir-Simpson scale for hurricane strength is based on maximum sustained wind speed, with Category 1 storms having at least 119 km/h (74 mph) winds, and Category 2 storms having at least 154 km/hr (96 mph) winds (Table 7.1). Sandy was a Category 2 hurricane when it was offshore Cuba on October 25, 2012, with maximum sustained winds of 175 km/hr (110 mph). However, it had been downgraded to a Category 1 hurricane by the time it made landfall in New Jersey, with maximum sustained winds of 150 km/hr (90 mph), and a maximum storm surge of less than 3 meters (10 feet). Hurricane and typhoon-related storm surge can reach 8 or 9 meters (26–30 feet), so Sandy was actually quite modest in size. Even so, it caused significant damage, especially along coastal New York and New Jersey, primarily due to storm surge-related flooding. Sandy is an excellent illustration of the vulnerability of much of the US Eastern Seaboard to flooding from storms and hurricanes.

Hurricanes are often considered purely tropical phenomena. Some weather forecasters used the rather clumsy name "extra-tropical cyclone" to describe Sandy (the media dubbed it Superstorm Sandy because it merged with a winter storm, affecting its track). Hurricane growth is determined in part by the presence of warm water at the ocean surface. Ocean surface temperatures in the North Atlantic have been increasing for the last few decades (Figure 7.11). Ocean surface temperatures in late summer and fall offshore northeastern US states and the Canadian province of Nova Scotia are now entering the range where hurricanes can be maintained and even strengthened well beyond tropical latitudes (Sandy should be considered the "new normal"). Urban planners, engineers and others involved in rebuilding efforts after Sandy in New York and New Jersey would do well to plan for more such events and to imagine what might happen when a Category 3 or 4 storm hits the area with storm surge of 5–6 meters (16–20 feet), instead of Sandy's mild Category 1 winds and waves and modest ten-foot storm surge.

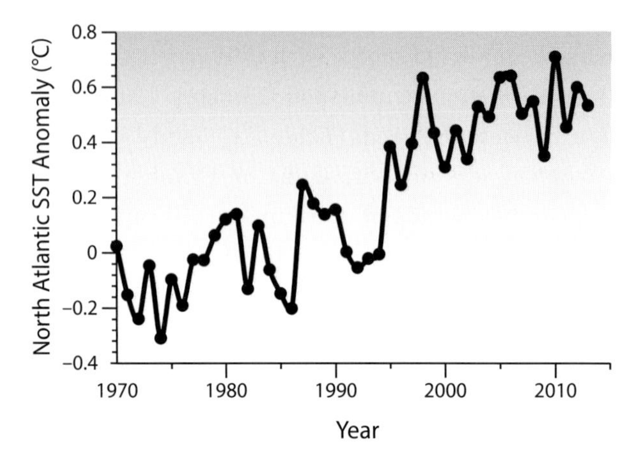


FIGURE 7.11 Average sea surface temperature (SST) in the North Atlantic since 1970, relative to the average temperature from 1901–1970. Modified from Yang et al. [2016].

## SUMMARY

Sea-level rise and future increases in storm intensity threaten our coastal infrastructure. Nuclear power plants are a good example. Many of these were built in coastal areas to take advantage of sea water cooling. Some face risks similar to those that overwhelmed TEPCO's Fukushima Daichi plant, as described in Chapter 4. TEPCO's backup power systems flooded when tsunami waves overtopped sea walls. The resulting cooling system failures led to overheating of nuclear fuel, core meltdowns, and release of radiation. A smarter engineering practice is to place backup generators and their control systems out of harm's way, for example on the third or fourth story of buildings where they will be safe from floodwaters. Coastal nuclear plants on the US Eastern Seaboard already face problems similar to TEPCO's during hurricane season – and these risks will increase. While nuclear plant licenses are only issued for several decades, once a plant is built, NIMBY (not in my back yard) politics almost guarantees that future plants will be built in the same location. Even if a plant closes, its nuclear waste will likely remain onsite, requiring residual cooling far into the future. Nuclear waste is difficult to ship, and in the US, there is no place to ship it to because Nevada is not accepting waste at its Yucca Mountain repository. Current nuclear plants or their waste storage facilities could be in the same location a century from now, so sea-level rise must become a major factor in our long-term planning.

Coastal nuclear power plants are likely to be part of our energy mix for many decades. We need a way to safeguard them in the face of rising sea level. These plants should be regularly stress-tested for flood-related failures, using storm surge models that account for sea-level rise, coastal subsidence, offshore bathymetry, and increased storm intensity. Armed with these models, engineers can design "bulletproof" systems for worst-case events, with double and triple redundancy for backup power, core cooling, waste cooling, and control systems. The Fukushima disaster has already prompted a reassessment of safety at US nuclear plants. To my knowledge, plants on the US Eastern Seaboard weathered Hurricane Sandy quite well. But nuclear planners also need to consider future trends. Coastal nuclear plants in no danger from earthquake-related tsunamis should plan for future flood levels that are 15 meters (nearly 50 feet) above present sea level: 2 meters of sea-level rise (a worst case estimate for a century from now) plus 10 meters of storm surge, plus an extra 3 meters for safety, for a total of 15 meters. Coastal plants in tsunami-prone areas, like Japan, need to plan for 40-meter flood levels. Figure 7.10 makes clear that high water levels from storm surge can remain for many hours. Facilities must be designed accordingly.

We can't predict the future exactly, but we do know certain things: World population is rising, as is population density in coastal zones. Sea level is rising, and storm intensity is increasing. These trends make clear that flooding will increasingly threaten crowded coastal areas. Coastal infrastructure must be better designed and built. This can be done at modest incremental cost if phased in over time, and if some of the lowest-lying areas are allowed to revert to a natural state. Glass and Pilkey (2013) describe an attempt by the legislature in North Carolina, a coastal US state, to preclude the use of realistic sea-level rise projections in state planning. In 2015, Pilkey was prevented from giving a lecture on sea-level rise at a state facility in Florida; the current governor, with the tacit support of the state legislature, discourages discussion of topics related to global warming and sea-level rise.

The governments of North Carolina and Florida are presumably unfamiliar with the legend of King Canute, an 11th-century Anglo-Scandinavian king who commanded the tides to stop, demonstrating

FIGURE 7.12 King Canute commands the tides to stop. From a 19th-century drawing by French artist Alphonse-Marie-Adolphe de Neuville.

to his sycophantic courtiers that even kings have limited power over nature. Attempts to "outlaw sea-level rise" in North Carolina and Florida would be comical if they didn't have such profound implications for future costs and risks for the states' people and their descendants.

The year 2013 was the 500th anniversary of Machiavelli's *The Prince*, with its common sense advice on politics and flood mitigation. Given the examples of Katrina in 2005, Fukushima in 2011, Sandy in 2012, and present knowledge of future sea-level rise, let's all resolve to do a better job avoiding future flood-related catastrophes.

# 8 What's All the Fuss about Global Warming?

People who like to garden in cooler climates often employ a greenhouse, composed of a glassed-in area where they can start their tomatoes and other vegetables earlier in the spring than if they planted them outside. It works because the glass keeps in the heat from the sun, even if it's cold outside. The glass allows passage of visible radiation from the sun (that's why we use glass for windows), which warms the earth, which in turn warms the air above the earth. The glass prevents the warm air from blowing away. Carbon dioxide ($CO_2$) plays a similar but not identical role in our atmosphere. Molecules like carbon dioxide and methane ($CH_4$) absorb heat energy that ultimately comes from the sun (some of it re-radiated from the Earth's surface) rather than letting it get reflected back to space. Without this greenhouse effect, the Earth would be intolerably cold. Too much $CO_2$, and the Earth would be intolerably hot. Right now we're close to that Goldilocks sweet spot where it is "just right" for humans.

The greenhouse effect and global warming are now accepted scientific wisdom. Yet, they remain controversial topics in many countries, with little consensus on what to do about them. Type the title of this chapter into an Internet search engine, and you'll find hundreds of entries, most of them skeptical of human-induced climate change and global warming.

In this chapter, I'll describe some of the background science, then discuss global warming in terms of the book's three main themes: communication, long-term processes, and economic impact. I'll show how the long-term aspects of the problem and various time lags in the system make it harder to communicate the science and implement solutions. I'll end with a few observations from Greenland, a place of stark beauty that is rapidly changing and holds important lessons for the rest of our planet. "Exercises for Students" in the Appendix provide further background.

When I was a graduate student in the 1970s, I attended a lecture by Charles Keeling, a scientist at Scripps Institution of Oceanography in La Jolla, California, on atmospheric $CO_2$ and its role in climate change. Keeling made the first reliable, multi-year measurements of atmospheric $CO_2$, using an observatory on Mauna Loa, far from anthropogenic (human) sources of $CO_2$. He initially took measurements in Antarctica, but it was too expensive to service the instruments; Mauna Loa was his second choice. Keeling's landmark studies, first published in 1960, clearly showed that $CO_2$ had annual cycles associated with Northern Hemisphere winter and summer. Plant growth in the summer consumes a lot of $CO_2$, which is released the following fall and winter when the leaves die. This phenomenon is controlled by seasons in the Northern Hemisphere, because the north has more land and, therefore, more plants than the Southern Hemisphere. Keeling's data would eventually show that average $CO_2$ had risen by amounts well above this annual variation. He was also able to demonstrate that this was due to burning of fossil fuels like coal and oil.

What's remarkable about Keeling's 1960 paper is that with only three years of data, he suggested that $CO_2$ in the atmosphere was increasing. He stated, "Where data extend beyond one year, averages for the second year are higher than the first year. At the south pole, where the longest record exists, the concentration has increased at the rate of about 1.3 ppm per year." Most people would not see the long-term trend in Figure 8.1 or grasp its significance. In fact many scientists were not convinced that a long-term trend actually existed. But by 1970, that trend clearly exceeded the annual variation, and the notion that global $CO_2$ was increasing began to be taken seriously.

A plot of $CO_2$ versus time is now known as the Keeling curve. Figure 8.2 shows data up to 2014. Over this longer time scale,

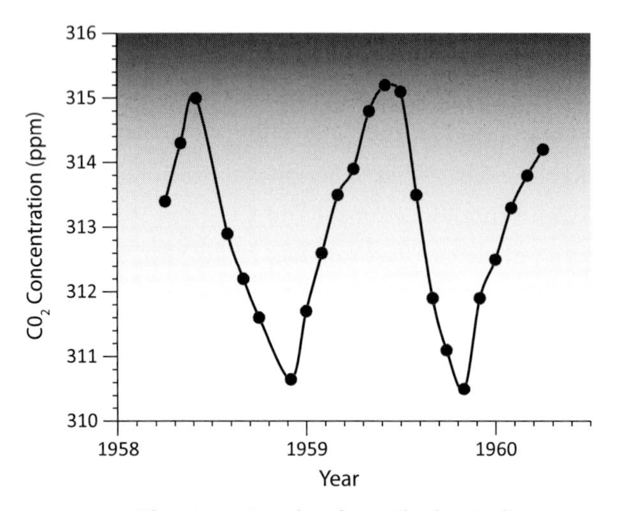


FIGURE 8.1 The Mauna Loa data from Charles Keeling's 1960 paper showing annual variation of atmospheric $CO_2$ concentration (in parts per million, ppm) from plant growth in the Northern Hemisphere. The sampling (almost monthly) captures the annual signal (compare to Figure A.3.1). Although Keeling postulated that $CO_2$ was increasing (based on limited additional data from Antarctica), longer-term observations would be needed before the increasing trend became apparent.

the long-term increase is obvious, as it's much greater than the annual change emphasized by Figure 8.1. The two curves together nicely illustrate the concept of adequate temporal sampling first highlighted in Chapter 3: good short-term sampling to describe annual or other "short-term" wiggles, coupled with a long enough time span to show the trend. For atmospheric $CO_2$, a sufficiently long time span is one or two decades. For earthquakes, it could be several thousand years.

It's worth emphasizing that much of the $CO_2$ we put into the atmosphere actually winds up dissolving into the ocean. Without this mechanism, air temperatures would already be a lot hotter. The downside is that the extra $CO_2$ makes the ocean more acidic. This affects marine life, especially species that use calcium carbonate ($CaCO_3$) to make their shells. $CaCO_3$ dissolves if the water is too acidic. Eggshells consist of $CaCO_3$ and will slowly dissolve in weak

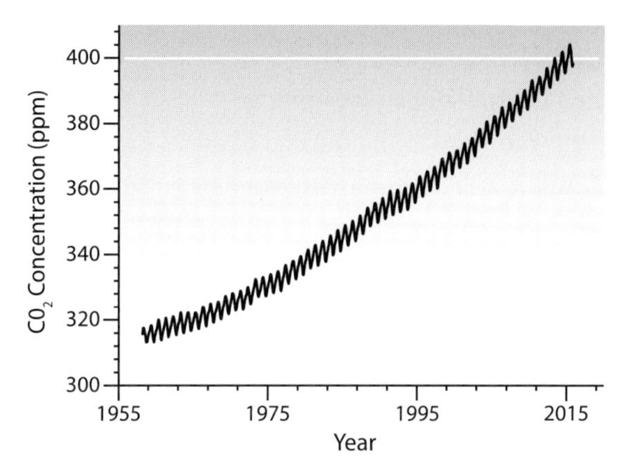


FIGURE 8.2 The Keeling curve shows $CO_2$ concentration in the atmosphere as a function of time. The small wiggles show the yearly fluctuation associated with Northern Hemisphere "greening" that begins in the spring (taking up $CO_2$) and ending in the fall (releasing $CO_2$), first shown by Keeling (Figure 8.1). The longer time span of this figure compared to Figure 8.1 clearly shows the trend of rising $CO_2$. The rate of rise is also increasing. When Keeling began his measurements, concentrations were rising at about 1.3 ppm (parts per million) per year. Now they are rising at 2 ppm per yr. Pre-industrial $CO_2$ levels are believed to be about 260–280 ppm. In 2013, we briefly hit 400 ppm (white horizontal line) for the first time in several hundred thousand years. We now routinely exceed this value. Figure by Theresa Maye, using data from Scripps Institution of Oceanography (http://scrippsco2.ucsd.edu/).

acids such as vinegar. You can try it at home: Put an egg in a cup of vinegar, and watch it slowly bubble away (don't leave the egg in the vinegar for more than a few minutes, otherwise the shell will become too thin and the egg will taste like vinegar). Marine species likely to be affected include corals, which use $CaCO_3$ to build their rigid framework, and some species of plankton that use $CaCO_3$ for protective shells. These small organisms form the base of the marine food chain. Lose them, and we risk losing higher order forms of marine life (big fish) that depend on them for food. This is such an important issue that the Schmidt Family Foundation, funded by former Google CEO Eric Schmidt and Wendy Schmidt, sponsors an X-Prize for the first person or group to develop better sensors to measure ocean acidity. Current designs need improvement – they are expensive and prone to drift and clipping (see Box 3.1).

The science behind global warming goes back a long way. Key parts of the theory were in place by the end of the 19th Century. In the 1820s, Joseph Fourier, a French mathematician, published a series of papers on heat transfer, and first pointed out that the Earth must have some sort of heat trapping mechanism to maintain equitable temperatures. Irish physicist John Tyndall measured the ability of $CO_2$, water vapor and other gases to absorb infrared radiation (heat energy) in 1859. Tyndall used a device similar in some ways to the instrument used nearly 100 years later by Keeling. In 1896 Svante Arrhenius published a paper showing how the Earth would heat up with continued burning of coal and consequent release of $CO_2$ (oil was not yet in widespread use). He won the Nobel Prize for this work in 1903. Keeling's continued measurements of rising $CO_2$ in the atmosphere in the 1960s and 1970s confirmed the basic theory. Since then detailed predictions of Earth's long-term climate response and shorter-term weather changes in specific regions have been modeled. What we should do about it and how much it will all cost remain significant challenges and are legitimate topics for debate, but the background science should not be in question.

Keeling reiterated a point made 70 years earlier by Arrhenius, that the increases in $CO_2$ being observed would inevitably lead to global warming because of the increased greenhouse effect. With the arrogance of youth, I remember thinking toward the end of his lecture that this was all pretty obvious, was unlikely to become a significant future research topic, and would certainly not be controversial.

Today, global warming is probably the most challenging environmental and public policy issue in the world. That this issue has become so contentious, and that there are so many skeptics and global-warming deniers, arguably ranks as one of the great communication failures in modern science. US Senator James Inhofe from Oklahoma has even called global warming a hoax. His beliefs, and those of many of his followers, are sincerely held. But to many scientists, this is like believing that the Earth is flat or that evolution is unproven. As scientists, we need to ask ourselves why so many people are skeptical of this important planetary-scale process, something that is scientifically accepted, and something that in the long run will affect the survivability of many species and key parts of the planet's ecosystems. It will certainly affect the health and economic well-being of our grandchildren and their grandchildren, so you might think there would be broad support for action. There is not.

How did we get to this sorry state of affairs? Like most complex topics, there are many reasons and relevant issues that could be discussed. I'll focus my discussion on this book's three themes: communication (including the need for the audience to listen), long-term aspects, and economic consequences. The Appendix has additional background and exercises for students.

## COMMUNICATION: THE NEED FOR LONG-TERM DATA AND THE CHALLENGE OF EXPLAINING THEM

Every spring, citizens of Alaska participate in a lottery known as the Nenana Ice Classic, named for the town of Nenana, on the Tanana River. Locals bet on when the ice will first move on the river during spring thaw. A tripod is placed on the ice, connected to a clock on shore. When the tripod falls through the ice or begins to move downriver, the clock is triggered, and the person guessing the closest time wins. The winning date is usually between April 20 and May 20. The contest has been run every year since 1917.

Figure 8.3 shows a plot of the winning date each year since the contest began, and a best fit line through the data. The plot is a good illustration of several climate change issues. First, a slight warming trend is apparent, with "ice-off" dates generally getting earlier over the one century the contest has run. But the amount of change over this period is actually quite small, only about a week-and-a-half in 100 years. Also, the scatter of dates is quite large. A typical person participating in the contest over his or her adult lifetime would not likely feel any warming trend or notice the ice-off date getting

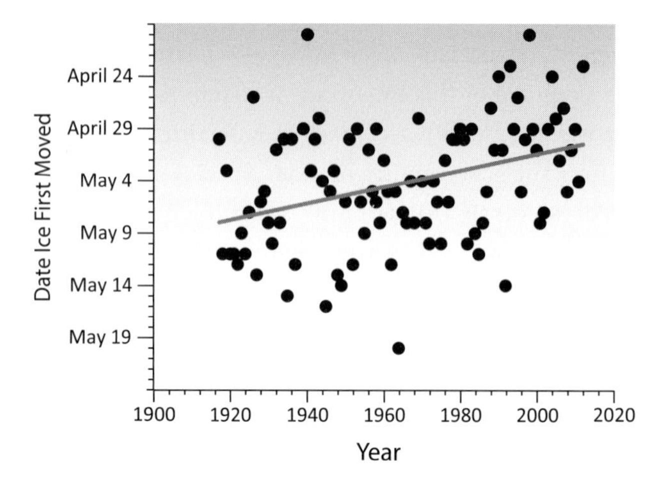


FIGURE 8.3 "Ice-off" dates on the Tanana River, Alaska, as a function of year, awarded in the Nenana Ice Classic contest, held every year since 1917. The best fit line through the data is also shown, and shows a small (~10 day) change over this period.

earlier: The change over two or three decades is too small. It certainly has little predictive value if you're trying to win the contest. Only when we plot up all the data over the full time span does a trend emerge, and even then it's subtle. The average Alaskan citizen could be forgiven if she or he was shown this plot and decided that the case for global warming was weak. As with the earthquake examples discussed earlier in the book, data spanning a longer period are needed to see a clear trend. Figure 8.3 illustrates one of the problems faced by climate scientists in communicating with the public – over one person's lifetime, its difficult to feel or "see" global warming.

Since good thermometers and data recording systems only became common about 100 to150 years ago, and good global coverage started even later, climate scientists have had to be clever at piecing together the planet's temperature change on longer time scales, just as geologists had to be clever to piece together the earthquake record. Figure 8.4 shows temperature estimates spanning the last 250 years. There is still a lot of year-to-year variability, and things are a bit "noisy" early on when there isn't much data, but the record in the

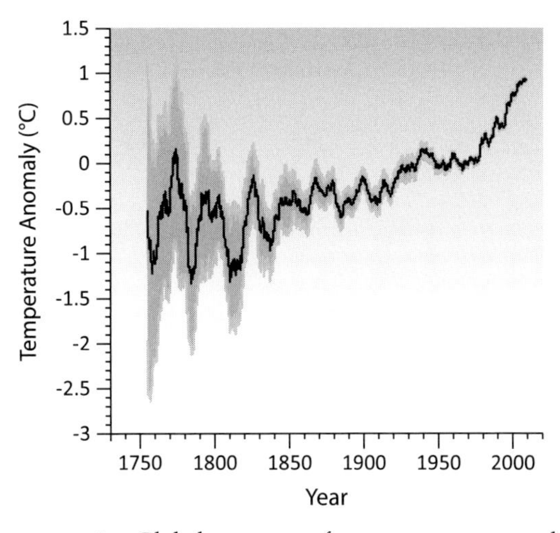


FIGURE 8.4 Global average surface temperature over land areas for the last 250 years, compiled by the Berkeley Earth Surface Temperature project.

last 100 years is pretty clear. Recent warming starts in the early 1900s, followed by a slight cooling between 1940 and 1970. After that, the warming trend is obvious, with temperatures consistently above their 250-year average. Figure 8.4 makes a better case for global warming than Figure 8.3 – the extra 150 years, and the global nature of the data, makes all the difference. Can we go back even longer, and make an even stronger case?

Dr. Michael Mann, a climate scientist at Penn State, has plotted historical temperatures going back a thousand years or more (Figure 8.5). No one can measure temperatures directly over such a long period, so indirect indicators ("proxies") have to be used, which involves making some assumptions. But lots of research by many groups has confirmed these assumptions. On these longer time scales, the recent rise in temperatures is even more striking, leaving no doubt about how much our planet has warmed in the last century. Figures 8.3, 8.4, and 8.5 together illustrate both the concept and the communication challenge inherent in global warming: It's only apparent over time spans longer than a human lifetime.

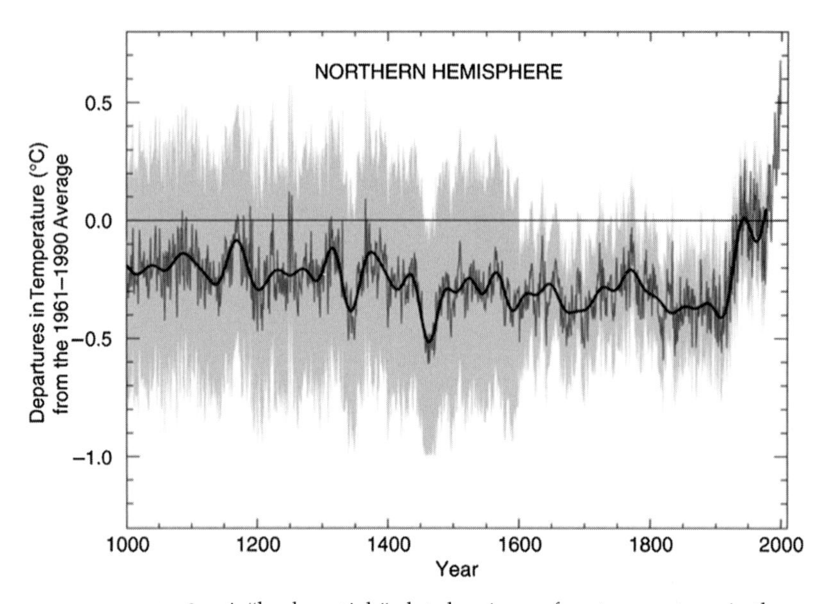


FIGURE 8.5 A "hockey stick" plot showing surface temperatures in the Northern Hemisphere for the last 1000 years. Gray band shows amount of uncertainty. Data for last ~150 years come from direct temperature measurements, such as those in Figure 8.4, with less uncertainty, while earlier data are estimates based on studies of corals, tree rings, ice cores, and other proxy records, and have more uncertainty. A period of overlap for some of the proxies with modern temperature measurements allows calibration. Mann et al. (2008) provide updated and more detailed versions of these data. Source: IPCC (Intergovernmental Panel on Climate Change).

Moreover, to get that long time span requires procedures not easily explained to non-scientists. Two of this book's main themes (communication, and the need for long-term thinking) are closely coupled when it comes to global warming.

As with Patterson and Needleman who fought to reduce lead in the environment (Chapter 6), vested interests savaged Mann's efforts and labored to debunk his findings, sometimes in unethical ways. His book on this topic (*The Hockey Stick and the Climate Wars*) makes for fascinating reading.

Unfortunately, many people are not familiar with the data or the methods that went into constructing Figures 8.4 or 8.5, or the long-term implications for the planet's health. A survey of public opinion in the US published in October 2015 indicates that the majority of the population remains skeptical of the concept of human-induced global warming. Only about a third of survey respondents believe that humans are the primary cause. Surveys in other countries suggest that even if the concept of human-induced global warming is accepted, it is not considered a significant problem. While all countries attending the UN-sponsored climate summit in Paris in 2015 agreed in principle to limit future $CO_2$ emissions, the gap between public attitudes and accepted scientific wisdom will make it difficult for governments to implement solutions, especially if those solutions cost money.

Some of the skepticism about global warming is related to something I call the good-driver syndrome. Most of us are convinced that our driving skills are above average, allowing us to safely barrel down a freeway at high speed (my driving skills are certainly above average, unless you believe my wife). Of course, we all cannot be above average. If you could take a one-week course on how to drive a NASCAR or Formula One vehicle at high speed, it would probably convince you that your driving skills prior to taking the course were actually not that good. In effect, our lack of knowledge about what it really takes to drive safely at high speed allows us to believe in the illusion that we are good drivers. Perhaps this is a trait bequeathed to us by evolution (the optimism gene?), whereby those with a confidant outlook on life tend to do better than the worrywarts. By the same token, for people who haven't learned very much about the science behind global warming, it's easy to dismiss the claims, especially if those claims fly in the face of everyday experience (think of those cold Alaskan winters). Charles Darwin, the famous English naturalist, summed it up nicely: "Ignorance more frequently begets confidence than does knowledge."

How could this general lack of knowledge about our planet become so widespread?

Like those of us who investigated earthquake risk in Hispaniola in the 1990s, or studied subsidence and low elevations in New Orleans prior to Hurricane Katrina, or studied giant tsunamis of the past in northern Japan, scientists have understood the research aspects of the problem for a long time, but were not able to get the message out. The public, the media, politicians, and government officials share some of the responsibility by not being very good listeners. Some of this "selective listening" likely reflects the influence of vested interests, but there are other issues as well. In the next sections I'll highlight some missteps made by both sides of the communication "fence," propose some solutions, and argue that things are slowly getting better.

### *Scientists as Poor Communicators*

The current public policy statement on global warming from the American Geophysical Union (AGU) is a model of clarity (full disclosure: I am a member of AGU and have served on its governance committees). The first line of the 2012 statement is "The Earth's climate is now clearly out of balance and is warming."

The reader has no trouble understanding where the organization stands and can read additional details if they choose by reading the rest of the statement. This statement follows the advice given to me by my sixth-grade teacher in composition class: "Always have a clear topic sentence."

The 2013 statement is even better – the title is the message: "Human-induced climate change requires urgent action."

Unfortunately things were not always so clear. Here is an earlier statement by the same organization from 2002, typical of many statements issued by scientists around that time: "Atmospheric concentrations of carbon dioxide and other greenhouse gases have substantially increased as a consequence of fossil fuel combustion and other human activities. These elevated concentrations of greenhouse gases are predicted to persist in the atmosphere for times ranging to thousands of years."

The first two sentences are clear enough, but they lack punch and also don't say anything about warming. The statement goes on to give a rigorous description of the status of scientific understanding of global warming at that time, with all of its complexity. However, it also goes on for 600 words, a clear violation of Rule # 1 in *Elements of Style* ("Use fewer words").

The 2002 statement ends with the following summary sentence: "AGU believes that the present level of scientific uncertainty does not justify inaction in the mitigation of human induced climate change and/or the adaptation to it."

The average member of the public or the media might be forgiven if he or she had difficulty understanding where the organization actually stood on the issue (some members did as well!). A careful reader of this sentence would note the double negative and might be able to reconstruct the logic to determine that AGU actually believed that something should be done about global warming. Note that the sentence begins by emphasizing uncertainty, a sure way to get the public to tune out and give ammunition to skeptics.

It is safe to say that AGU's 2002 statement, and similar ones from other scientific societies at the time or earlier, did not have a significant impact on public policy or public perceptions of climate change. Fortunately, as we have seen, later statements by AGU improved significantly. But the earlier statement is typical of many made by scientists in the 1980s, 1990s, and early 2000s, and is still occasionally heard today. Even if today's scientific organizations have improved their message, the present state of public confusion, skepticism, and public policy malaise reflects, at least in part, several prior decades of poor science communication.

### *Communicating Uncertainty*

Recall again the first part of the summary sentence in AGU's 2002 statement: "AGU believes that the present level of scientific uncertainty ... " The statement emphasizes uncertainty, rather than the more important thing – what we know. When speaking to the media, most scientists will first discuss the uncertainties of their measurements, then describe the complexity of the problem, and finally counsel caution in the interpretation of their results. That approach is hardwired into most of us when publishing a scientific paper. I am as guilty as other scientists. One of my frequently cited papers, written with my former graduate student Dr. Ailin Mao and my colleague Dr. Chris Harrison, focused on quantifying the uncertainties of measuring plate motion and earthquake-related deformation with high precision GPS. Although many scientists have used this work in their research, I'm pretty sure the average member of the public would have no interest in it. While the emphasis on uncertainty in scientific measurement may be endemic in the scientific community, focusing on uncertainty does not work when speaking to the media, the public or government officials. Instead, we need to give a clear, simple message: Given increasing $CO_2$ in the atmosphere from fossil fuel burning and our knowledge of how a greenhouse atmosphere works, the planet must heat up.

The scientific community has known this since 1896, but it is rarely stated clearly. After a particular heat wave, drought, or storm, many scientists, when interviewed by the media about a possible connection to global warming, will first say that it's difficult to prove that any one event is directly attributable to global warming and then go on to give a scientifically rigorous statement about possible statistical correlations, with appropriate caveats, an explanation of uncertainty, and perhaps a short introduction to radiative transfer theory. But it doesn't matter much what these scientists say after that initial "It's difficult to prove" statement, because the public has already tuned out. What they probably heard was "It's not global warming."

As scientists, we need to get better at giving clear, simple statements, without all the usual obfuscation. When asked if there is a connection between a particular heat wave and global warming, why not just say "Yes"?

To skeptics (and many scientists) who say that it's difficult to attribute an individual weather event to a long-term trend like global warming, I am tempted to say "If it walks like a duck, and sounds like a duck, its probably a duck." However, more statistically rigorous arguments to this effect can also be made. See, for example, recent publications by Rahmstorf and Coumou (2011) and Hansen et al. (2012). Hansen is a well-known climate scientist, recently retired from the Goddard Institute of Space Studies, New York. In their 2012 article, Hansen and colleagues point out that in a warming world, not only will the average temperature increase, but also the number of extreme heat waves. Both groups of authors go on to point out that the chance that recent record heat waves are unrelated to global warming is extremely small.

I'm going to illustrate this point with a simple thought experiment. Let's imagine we are in some mid-latitude location in the Northern Hemisphere, and measure morning temperature each day during one summer month. We get 31 measurements between 10°C and 30°C (50°F and 86°F), with an average close to 20°C. If we plot these on a histogram (Figure 8.6a), which is just a way of showing how often certain values occur, we see a peak of values near 20°C (the average, or what scientists like to call the "mean"). Nine days were between 20°C and 22.5°C. The width of the histogram shows the spread of values. One measure of spread used by statisticians is something called the standard deviation, which in this case is about 4°C. This means that about two-thirds of our 31 temperature measurements lie between 16°C (20°– 4°) and 24° (20°+ 4°). If we added a lot more temperature measurements to the plot, the histogram would become taller and smoother, but its basic shape, which statisticians call a bell curve, would stay the same.

Now imagine that instead of just one person doing these measurements in one place for one month, we are able to recruit an army of amateur scientists who do these measurements in hundreds of locations across a larger region for three summer months for 10 years. Instead of 31 measurements, we have hundreds of thousands of measurements and a very good description of the average summer weather in a certain region for that decade; we should get a pretty

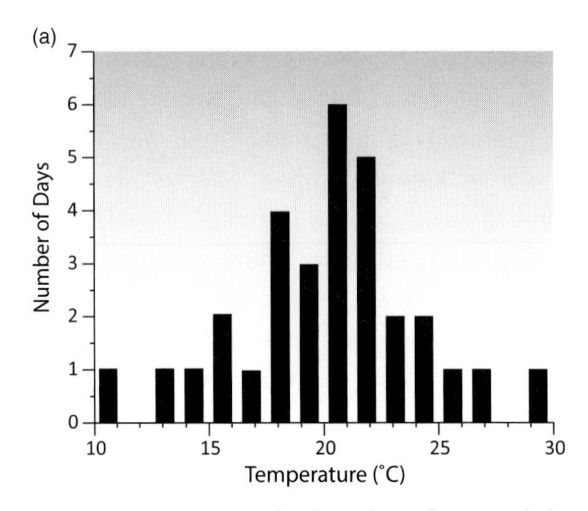


FIGURE 8.6A Histogram of 31 hypothetical average daily temperature measurements taken over one summer month.

smooth bell curve. Now lets imagine that we have that kind of data in an earlier decade (e.g. 1951–1960) and a more recent decade (e.g. 2001–2010) (Figure 8.6b). The peak of the bell curve (the average temperature) shifts to the right in the more recent decade compared to the earlier decade, indicating warmer temperatures. The way I've drawn it, the spread (or standard deviation) has also increased slightly, indicating more extremes in both hot and cold temperatures. Similar diagrams apply to annual temperatures. Because the more recent decade has shifted to a warmer average, it should be pretty obvious that we're going to see more record summer heat waves compared to record winter cold spells. There could still be some pretty cold winters, just fewer of them. The implication is that there are going to be a lot more events like the 2003 heat wave in Western Europe, the 2010 heat wave in Moscow, and the 2012 heat wave and drought in the US Midwest, especially in the central parts of continents that are far from the moderating effect of the ocean.

To see a real world example of this kind of statistical argument, lets look more closely at the 2012 drought and heat wave in the US. Some climatologists think that the roots of this event actually started

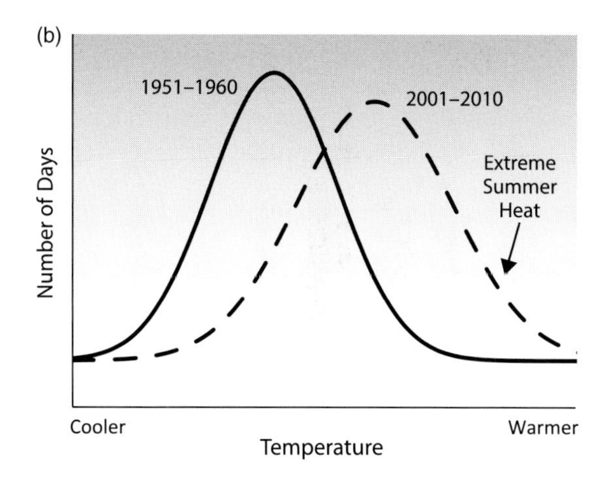


FIGURE 8.6B Hypothetical histogram of daily summer temperatures for a mid-latitude region, averaged over two separate decades. The more recent decade is warmer, increasing the chance of summer heat waves.

earlier in the 2011 drought and heat wave in Texas. Droughts and heat waves are often related, for several reasons. For example, when it's really hot, evaporation is faster, and plants quickly become water-stressed. Also, hot air is farther from the dew point (the cool temperature where atmospheric water vapor starts to condense into rain drops), so it's less likely to rain, at least over dry continental interiors (over oceans, you get more evaporation, and it can actually rain more). It should come as no surprise that when its gets hot in Texas in the summer, there's less rain.

Figure 8.7 illustrates this point. It shows summer temperature and summer rainfall in Texas since 1895. The cloud of data is an indication of background variability during this 115-year period. While there is some scatter, it's easy to see the overall trend, shown by the straight line – high temperature usually means less rainfall. Now take a look at the "wagon wheel" in the upper left – that represents the summer of 2011. It's a lot hotter and a lot drier than any previous year on the plot. You don't need a Ph.D. in statistics to see that something radical occurred in 2011. That summer in Texas was an extreme event, breaking all previous records. The 2011 drought

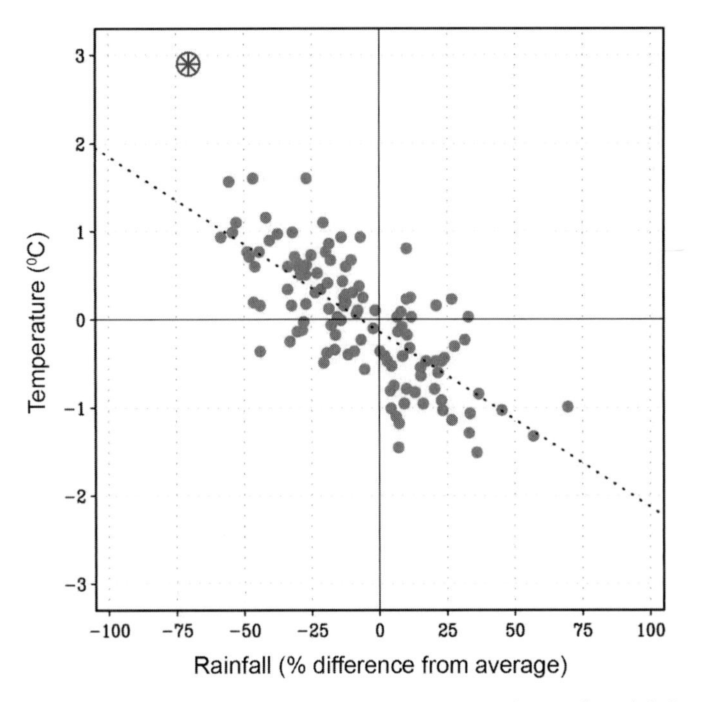


FIGURE 8.7 Average summer temperature in Texas (vertical axis) defined by June–August temperature anomaly, or difference from average, compared to rainfall (horizontal axis) for every year between 1895 and 2010 (small circles). The anomalous 2011 summer event is shown by the "wagon wheel" on the upper left. The dotted line is the best fit line through the data and emphasizes that hotter summers tend to have less rainfall. Modified from Hoerling et al. (2013).

cost Texas approximately $8 billion, mainly in crop and livestock losses.

Most scientists, myself included, tend to use a lot of jargon when describing their science. Dr. Richard Somerville, a climate scientist at Scripps Institution of Oceanography, and Susan Hassol, Director of Climate Communication, a nonprofit group, published a table in 2011 showing a list of terms often used by climate scientists and how these terms can be misinterpreted by the general public. Some of their entries are quite humorous, or at least they would be if the subject were not so serious. My favorite is "aerosol." This is a term scientists use to describe small particles in the atmosphere, which, especially at high altitude, affect Earth's albedo (how reflective it is) and how much solar radiation gets reflected back to space. However, most people interpret "aerosol" as "spray can," and then get global warming confused with the ozone hole.

One term that is not in Somerville and Hassol's table that perhaps could be is "model." To most people, a model is meant to represent a larger object like a ship or airplane, and it can be a flimsy construction of plastic and glue, or even Popsicle sticks. To Earth scientists, however, a model is one of the most powerful tools we have for figuring out how the Earth works. A simple one might be just a short equation, solved with pencil and paper or calculator, for example the equation in the caption to Figure A-7.1. The equation describes what we think the basic physics of the problem should be. It may not describe everything that happens, but it describes most of the important stuff. It makes a prediction about a certain physical variable that we can measure, such as motion near the San Andreas fault (Figure A-2.2) or sediment compaction (Figure A-7.1). We can then compare our measurements to the predictions of the model. If the measurements and model predictions agree, we are beginning to understand something useful about that particular problem. Results can be presented at a professional meeting, criticized, tested, and improved. Within in a few years, the scientific community should have a pretty good understanding of the process – the model becomes accepted, at least until a better one comes along.

These days, the models of climate scientists are far too sophisticated to be described by a simple equation or solved with pencil, paper, and calculator. Giant supercomputers run programs consisting of hundreds of thousands of lines of computer code, assimilating millions of measurements from all over the world, and adjusting hundreds or thousands of variables. These models are not flimsy constructs. In their own way, they are as sophisticated as the hardware, firmware, and software that discovered the Higgs boson, a fundamental building block of matter, at the Large Hadron Collider, a giant particle accelerator on the French-Swiss border.

When the climate models suggest trends of increasing summer heat waves, we should take them seriously, even if a specific summer in a specific place is not especially hot. Remember, the models and predictions are statistical in nature. But just as a gambler will surely lose in Las Vegas over the long run, the planet will surely heat up.

The very sophistication and complexity of climate models, and the physics they encompass, has made it challenging for climate scientists to describe their results to the public. The trees get in the way of the forest. How, then, to communicate the science? Perhaps we should follow the example of the medical profession. Medical doctors usually skip the jargon and go straight to the point (for example, stop smoking, get more exercise, eat more vegetables). Patients who want to know the gory details can usually get them, but most of the time, people just want the bottom line. Pretty much the same prescription applies to the planet: We need to use less coal and oil (less smoking) and develop lifestyles and urban designs that favor walking for short trips instead of using the family car (more exercise). Eating more vegetables might help as well – not only is it good for our health, but meat production tends to generate more greenhouse gases than other types of food production.

I'd like to end this section on a positive note. Scientists are getting better at communication. The 2013 annual meeting of the American Association for the Advancement of Science (AAAS) had an entire day devoted to science communication; AAAS also funds fellowships in mass media, where young scientists spend a year or two working with media organizations. AGU now holds well-attended seminars to improve scientists' communication skills and has an annual prize for climate communication. Richard Somerville and Susan Hassol have started a nonprofit group called Climate Communication. Michael Mann and colleagues have started a website (www.realclimate.org/) that explains the science behind global warming, and debunks myths propagated by special interest groups. Non-scientists are also helping. A moving documentary on the world's shrinking glaciers by film maker James Balog was nominated for a 2013 Academy Award. These efforts and many more like them will eventually start to influence public discourse and policy, but it may take a while, as described in a later section (The Time Lag Problem).

### *The Public and Politicians as Poor Listeners: The Role of Media*

As with the other examples discussed in this book, responsibility for our present state of affairs reflects a combination of poor communication skills by scientists and poor listening skills on the part of the public, politicians, and the media. In other words, communication is a two-way street, and we all need to take some responsibility to make it work.

I have never heard a climate scientist state that global warming is going to do away with cold winters. Yet every time we have a cold winter or a good old-fashioned snowstorm somewhere on the planet, a climate change skeptic will announce that global warming is not real or that scientists are exaggerating the problem – or worse, that scientists are involved in a hoax. And various media outlets will uncritically report these statements. This has always baffled me. Most people understand the idea that you can have short-term fluctuations superimposed on longer-term trends, like cool nights and warm days during a gradually warming spring. Are skeptics and media pundits so lacking in imagination that they can't envision the occasional cold winter superimposed on a multiyear warming trend? It's as if people are forgetting their high school geography lessons on the difference between weather and climate, or those common-sense folk sayings. (My favorite: Weather determines the clothes you wear. Climate determines the clothes you buy. Both change.)

After a cold snap that affected much of the US in early January 2014, Rush Limbaugh, a well known personality on Fox News, a media outlet owned by Australian media tycoon Rupert

Murdoch, stated that "Obviously there is no melting of ice going on at the North Pole. If they're gonna tell us the polar vortex is responsible for this cold, that means record cold is also happening in the North Pole, which means there isn't any ice melting." Limbaugh was referring to the loss of Arctic sea ice. The loss has been going on for at least several decades and recently accelerated. It has big implications for Arctic ecosystems and is also a frightening indicator of major planetary change. Apparently Limbaugh forgot, or does not understand, that sea ice losses happen mainly in the summer, when it's warm. The long-term decline in Arctic sea ice mainly reflects warmer and longer Arctic summers, and it is not especially sensitive to the occasional mid-latitude winter cold snap. Of course, as with most things climate-related, the story is more complicated, but also more interesting.

### *The Sea Ice Story*

Arctic sea ice is a good indicator of the global warming issue and all its complexities. If you look at maps of Arctic sea ice spanning a decade or two, it would be difficult to separate yearly changes from longer term changes – the ice moves around a lot in response to changing winds and currents. There is significant year-to-year variability, and huge winter-summer differences: *When* you look affects *what* you see. There are also measurement challenges. Sea ice is difficult to measure without satellites (ships can't travel there, except for brief periods in the summer), and we only have good satellite measurements starting in the late 1970s or early 1980s. So we lack a good long-term record. There is even a definition problem: Should we look at the total area of ice, including open ocean areas with some floating ice (scientists call this sea ice extent) or just the part that is completely frozen (scientists call this sea ice area)? Figure 8.8a shows sea ice extent, for most of the years that good satellite data are available. The summer-winter variation is clear, the longer-term trend less so. Figure 8.8b zooms in on the summer melt season, when most losses occur. Now the long-term losses are more apparent. If we plotted all the data instead of just every

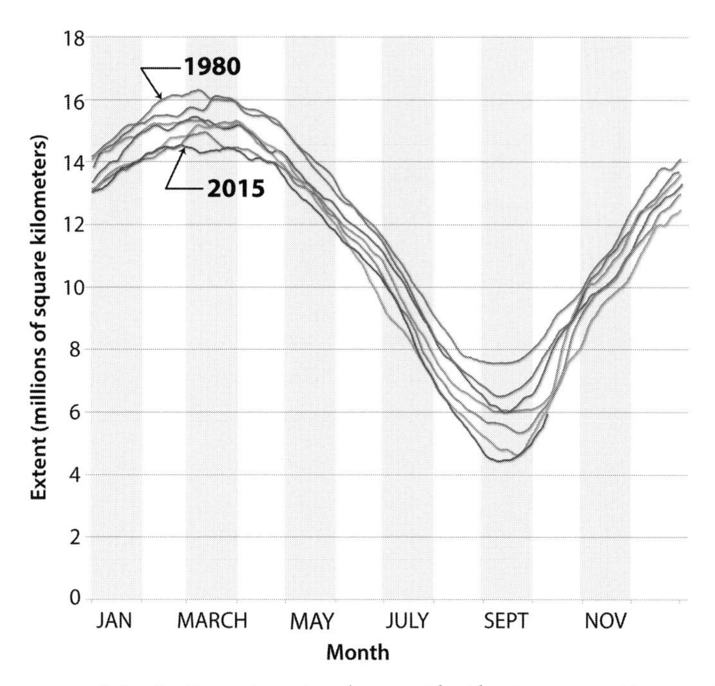


FIGURE 8.8A Arctic sea ice extent (areas with at least 15 percent ice cover) plotted every five years since 1980. The maximum occurs in late February or March at the end of winter, while the minimum occurs in September at the end of summer. Data courtesy of National Snow and Ice Data Center (NSIDC).

fifth year, the plot would be quite busy, like a bunch of spaghetti, but you would see that summer 2012 actually had the lowest area on record, at least as of 2015.

One number of interest not clear from these two graphs is when the Arctic will be ice-free in summer. This is an important date for shipping companies, who can save a lot of money by taking an Arctic "short cut." For that analysis, sea ice area is a better indicator, since ships can bypass small areas of floating ice. Figure 8.8c shows the minimum area for each summer plotted by year. Not only are we losing ice, but the losses have accelerated. We can estimate the year when this part of the ocean will be ice-free in summer by drawing a best-fit line through the data and extrapolating to zero – that exercise

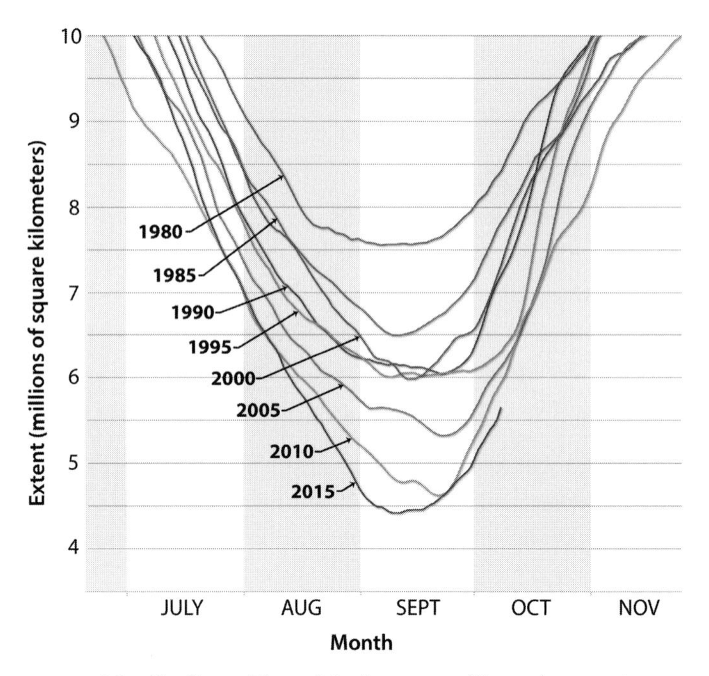


FIGURE 8.8B Similar to Figure 8.8a, but zoomed in to show sea ice extent in the summer. By summer 2015, sea ice extent was about two-thirds of the 1980 value.

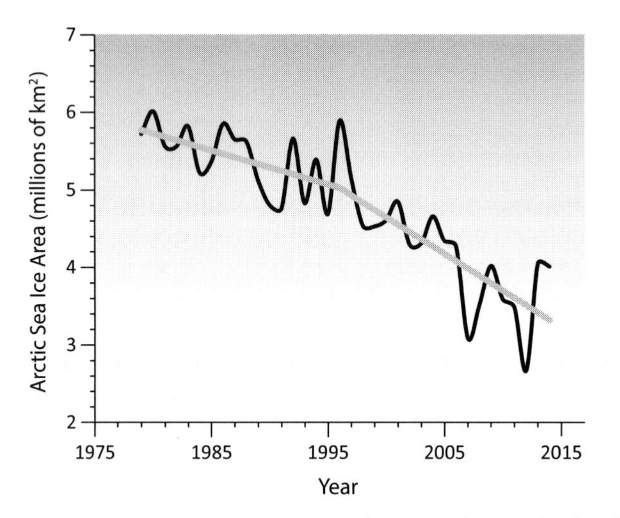


FIGURE 8.8C Arctic sea ice area as a function of time. The data (black line) show the summer minimum ice area for each year, and are fit with two straight lines (gray lines). The later one, from 1996 to 2014, shows summer sea ice decreasing faster than the previous two decades. If this trend continues, the Arctic will be ice-free in summer by 2050 or earlier. Modified from Yang et al. (2016).

suggests a date sometime between 2040 and 2050. Some analyses predict even earlier dates, and some shipping companies are already developing plans to use the summertime Arctic route.

One thing not shown by these graphs is the volume of ice (i.e. including ice thickness), which is actually more important but more difficult to measure. A relatively warm winter can set up the following summer for a big sea ice loss because less ice forms – and what does form is thinner and easier to melt the next summer.

The cold snap in early 2014 referred to by Limbaugh on his Fox News telecast affected the eastern US and southeastern Canada. On a couple of days in January, the city of Atlanta, Georgia (near 34° North latitude), was actually colder than Anchorage, Alaska (61° North latitude). As the cold Arctic air mass moved southeast across Canada and the US, warmer air from the Pacific Ocean moved northwest into the Arctic to take its place, warming many northern areas that would usually be pretty cold at that time of year. For the first time since records were kept, many parts of Alaska did not experience temperatures colder than 0°F during 2014. Some winter cold snaps that affect mid-latitudes in eastern North America might actually wind up reducing the formation and thickness of winter Arctic sea ice, making that ice more vulnerable to melting the next summer.

## *Media Obfuscation*

Most television or radio reports on climate change follow a standard formula. In a typical report, the author of a scientific study will have a few seconds to describe his or her work. This will be followed by an opposing view, often by someone with dubious scientific standing or lacking relevant credentials. Frequently the spokesperson giving the opposing view will be associated with a foundation funded by special interests that have a financial incentive to maintain the status quo (e.g. exploitation of fossil fuel), or a politician who has accepted campaign donations from special interest groups. All too often the journalist will fail to question these obvious conflicts of interest. This strikes me as sloppy journalism.

Here's an example. On March 27, 2013, National Public Radio (NPR), a major US media outlet, aired a story that discussed the difficulty science teachers were having in developing curricula for younger students on the carbon cycle, in part because of recent human interruption of this cycle and its implications for global warming ("A Hot Topic: Climate Change Coming to Classrooms"). Jennifer Ludden, the NPR reporter, aired an interview with Mark McCaffrey of the National Center for Science Education (NCSE). McCaffrey worked for a number of years at the Cooperative Institute for Research in Environmental Science, an institute at the University of Colorado, before joining NCSE and is an expert on educational issues related to the environment. Ludden then aired an opposing view from James Taylor, who is employed by the Heartland Institute. What Ludden failed to tell her listeners was that Taylor had no qualifications in either climate science or education (he had trained as a lawyer). While I have a lot of respect for lawyers, I don't think they are qualified to determine the science taught to our children. Moreover, Ludden failed to inform her listeners of the various sources of funding for either interviewee. Information on NCSE's finances are publicly available. It receives funding from its members (mainly science teachers) and supporting organizations, which include the Howards Hughes Medical Institute and the American Association for the Advancement of Science. The Heartland Institute is a privately funded think tank whose funding sources are not publicized (usually a bad sign). Several publicly available reports have suggested that fossil fuel interests are major contributors to the Heartland Institute. These reports should have raised red flags for Ludden. An obvious question to pose to Taylor during the interview would have been "Is your salary paid by people or organizations who would gain financially from continued use of fossil fuel?" Ludden failed to ask this obvious question.

This approach to climate change reporting reminds me of "teaching the controversy," a tactic developed by anti-science groups opposed to teaching evolution in schools. These groups believe in creationism. They want science teachers to present "both sides" of the evolution/creation story, as if Darwin's powerful theory of life, which underpins all of modern medicine, biology, genetics, and geology, should be weighed equally with the views of a small number of religious groups. For many years, tobacco and oil companies used similar tactics, attempting to raise doubts about the work of scientists looking at the health risks associated with smoking and leaded gasoline. It is disappointing that a respected news organization like NPR would succumb to the logical fallacy of "reporting the controversy." Mayer (2013) describes an especially troubling example of private money influencing public media.

A related issue: Why does the news media uncritically report the views of politicians on the science behind global warming, without even minimal follow up? In June 2012 in a television interview, Rick Santorum, a Republican candidate for the US presidential nomination, stated " ... the idea that man, through the production of $CO_2$ – which is a trace gas in the atmosphere, and the manmade part of that trace gas is itself a trace gas – is somehow responsible for climate change is, I think, just patently absurd." Santorum was correct that $CO_2$ is a trace gas, and his views reflect a common misconception, namely that because $CO_2$ concentrations seem low (a few hundred parts per million), it cannot cause much planetary heating. However, it's been known for a long time that even a trace gas can have a big impact when it comes to heat. John Tyndall pointed it out in a public lecture in 1863, based on his earlier experiments on the heat-absorbing properties of various gases. I don't expect candidates for public office to be experts on this. It's a difficult concept, one that beginning college science students often struggle with (a full explanation includes the arcane subjects of quantum mechanics and statistical mechanics). It's not surprising that non-scientists would have this misconception. But candidates for public office should at least have some respect for the views of scientists who have spent their careers studying the problem, and have tried to warn the public and policy makers of critical long-term implications.

Santorum's statement was unfortunately typical of statements by several candidates in recent US elections. If a politician's views on a scientific (or any other) topic are reported, why not give equal time to someone who is qualified to discuss that particular topic? If this practice became the norm in political reporting, voters might be surprised to learn how little candidates for public office actually know about any number of critical issues. If candidates for public office knew that immediately after they spoke there would be follow-up commentary by experts, it might make politicians more circumspect in their political speech. Cynthia Barnett, author of *Rain* (2015) has put it succinctly: "Will we be guided by science, or ... the influential uninformed?"

If your family doctor told you that you had a serious heart condition and recommended that you see a specialist, you would probably go to a qualified cardiologist to get her or his opinion on the correct course of action. If instead you chose to go to a politician who suggested that you ignore the problem, and you chose to listen to that politician instead of your doctor, there is a good chance that you would soon find yourself in serious trouble. By the same token, if we listen to skeptical politicians rather than qualified climate scientists in the climate change debate, there is a good chance our planet will be in serious trouble. The problem is that we will not feel the full effects for many decades. Our grandchildren and great grandchildren will be affected much more than us, because of time lags in the planet's response. Which brings us to the second theme of this book, the issues of time lag, time scale, and future implications. We've already seen that it takes a long-term record of past temperatures to verify that things are changing, and the challenges scientists face both in measuring this change and convincing the public that it's real. Now I'll consider future changes, and how actions over the next few decades can affect the liveability of the planet far into the future.

## THE TIME LAG PROBLEM: WHY WE NEED TO THINK LONG TERM

In 1993, an elite US military force was involved in a UN-sanctioned mission in Mogadishu, the capital of Somalia in northeast Africa.

The larger UN mission was to provide security for NGO's and other groups delivering food aid to Somalia, then in the grip of a deadly famine. The US force was attempting to arrest two lieutenants of a local warlord, Mohamed Farrah Aidid, who had been stealing food aid meant for the citizens of Somalia. He had also been killing or kidnapping aid workers. Despite an array of high-tech weaponry and supporting technology, the mission went badly wrong. A key problem was time lag, a well-known concept: We often use the phrase "out of sync" to describe situations where the timing of things is not quite right.

In the case of Mogadishu, US ground forces, with an armed convoy of Humvees and trucks, had to navigate into and out of the city center during the mission. The convoy was assisted by a surveillance aircraft equipped with high resolution video and a bird's eye view of the city. This aircraft would direct traffic, making sure the convoy could navigate the narrow streets of Mogadishu, avoiding ambushes, road blocks, and other potential choke points. The problem was that the radio frequency used by this aircraft was not the same frequency used by the ground vehicles. Radio messages had to be relayed through several intermediaries: first to the operations commander back at a base outside the city, then to a lower altitude command helicopter, and finally to the convoy. This introduced a delay of perhaps 10 or 15 seconds between the time the radio message was first sent from the surveillance aircraft and when it was finally received on the ground. If the convoy was approaching an intersection, and the surveillance airplane decided the convoy should make a turn, the time lag between transmission and reception meant that by the time the message was received, the lead vehicle had already missed the turn. It's difficult to back up a long convoy (especially if you're taking enemy fire), so the lead vehicle could either keep going, hoping for additional directions, or stop the convoy and try to get it turned around despite the narrow streets and attendant delays and risks. While the mission was ultimately successful, the resulting confusion and additional time spent in hostile territory contributed greatly to the number of casualties, ultimately leading to the pull-out of all US forces from Somalia.

We have a similar time lag problem with climate, but instead of 10 or 15 seconds, it's more like several hundred years. We've already put a lot of $CO_2$ into the atmosphere, but for most people, things don't feel so bad right now. The problem is that the full impact of this $CO_2$ won't be felt for a long time, making it difficult to communicate the problem or get support for concerted action. Kunstler (2006) coined the term "long emergency" which nicely describes our climate change dilemma. The implication is that things are going to get a lot worse before they get better. But we won't be around to feel it. Our great-grandchildren will be stuck with the bill.

Sea level is a good indicator of the time lag problem. The last time the Earth had 400 ppm $CO_2$ (the current value) in its atmosphere was three million years ago. At that time, sea level was at least 6 meters (nearly 20 feet) higher than today. While scientists are still debating the exact height of sea level three million years ago, and why it was so much higher than today, the simplest explanation is just time lag: 400 ppm $CO_2$ will eventually cause warming sufficient to melt some of the Greenland and Antarctic ice sheets and raise sea level by at least 6 meters. But it takes the ice sheets some time (probably a few hundred years) to respond and reach a new equilibrium state. One clue that this process has already started is that Greenland's mass loss recently began to accelerate (Figure 7.6). Because of time lags in the system, most of this sea level rise is already "locked in." In other words, even if we start to curtail our $CO_2$ emissions in the near future, we are probably stuck with at least 400–450 ppm $CO_2$ for the next few hundred years and, hence, could be stuck with at least 6 meters of sea-level rise. The economic consequences will be profound.

I don't know the best energy solutions that should be adopted in the future, but I do know there will also be a time lag related to transitioning to low carbon alternatives. The transition can't happen overnight – our industrial society has been carbon-based for too long a time. The industrial revolution began in the early 1800s, with the development and refinement of the Watt-Boulton steam engine. But we were dependent on carbon long before that: Burning wood was our primary way of cooking and keeping warm for thousands of years. Many early steam engines ran on wood. It's often said that wood is renewable, but that's only true if you replant the trees, not if the cleared forest is later used for crops or urbanization. My ancestors did not use wood renewably. Having stripped most of Europe of its original tree cover, they moved across the Atlantic and proceeded to do the same thing to North America's original old-growth forests. Most of what we see today in North America is second- or even third-growth forest. If you think the trees are the same, you are mistaken. Just go into an old barn, and look at the main beams holding it up. These are hardwoods, usually oak, elm, beech, or maple, from trees much larger than commonly seen today. Our economy has been based on cheap carbon – either wood, coal, oil, or natural gas – for a lot longer than a few hundred years. Even if a good alternative appears tomorrow, it will take at least several decades to re-engineer our transportation, electrical generation, and industrial infrastructure to exploit it.

We also have a time lag problem in communication and public attitudes. Fortunately this one is shorter and will likely go away soon enough. While current skeptical attitudes stem in part from poor communication in prior decades, scientists have become much better at getting the message out. It may take another decade or two for our improved communication skills to have much impact, but a few more summer heat waves and Superstorm Sandys may tilt the debate. Either way, I believe that within the next few decades there will be broad public acceptance of the idea of human-induced global warming. A strong consensus in favor of doing something about it, especially if it costs a lot of money, may take a few more decades. Combine the political time lag with the time to re-engineer our energy infrastructure and the time lag between $CO_2$ emissions and maximum heating, and you can see that our home planet is likely to be a very uncomfortable place for our grandchildren's grandchildren.

Many scientists and policy experts believe that we should limit warming to 2°C above pre-industrial levels. My guess is that we are already well past that point, but because of various time lags in the system, it hasn't shown up on the thermometer yet.

## ECONOMIC CONSEQUENCES

I'm sometimes asked by students what the future will look like when global warming starts to happen in earnest. I usually answer that we will see

1. More summer heat waves and more deaths from heat stroke, dehydration, and heat-related illnesses.
2. More droughts, especially in continental interiors.
3. Accelerating loss of sea ice, glaciers, and ice sheets.
4. Rising sea level.
5. More intense storms.
6. Increased coastal flooding.
7. Loss of marine species from ocean acidification.
8. Expanding range for tropical diseases.
9. Mass migration of people from low lying flood-prone areas to higher ground and from hot equatorial areas to cooler northern regions.

Each of these events, by themselves, has the potential to cause significant social and economic disruption. Taken together, the economic consequences will be profound. Astute readers will notice that most of these things are already happening. Some experts will argue that things are more complicated and that there is more than one explanation for these phenomena. While this is true, I think a case can be made that for each item on the list, global warming is a key underlying factor.

Consider point #9, perhaps the most controversial. Bangladesh, which struggles with land loss and resulting loss of economic opportunity associated with rising sea level, has already seen increased migration to nearby countries like India, Malaysia, and Indonesia. The US and Mexico are seeing increased migration from Central America, a region where several countries struggle with high levels of poverty and violence. October and November 2015 saw more than 10,000 unaccompanied children, most from Central America, cross the US-Mexico border. The youngest was six years old. The 2015 numbers were nearly double those of 2014. The parents believe their children stand a better chance of being accepted as refugees if they are alone. It is difficult to imagine the level of desperation that would compel parents to send their children away, alone, into an unknown land and unknown future.

There were also large increases in human migration from Saharan and sub-Saharan Africa to Europe in 2014 and 2015, especially through Italy and Spain. People were so desperate to leave that they were willing to cross the Mediterranean in small boats, risking their lives and the lives of their children. More people drowned in the Mediterranean from these crossings in 2014 and 2015 than died in the sinking of the Titanic. During a single week in May 2016, more than 1,000 migrants drowned trying to go from Libya to Italy. Other parts of Europe saw huge increases in migration from Syria during the same period, reflecting that country's brutal civil war. But even here, climate change may have been an underlying factor. Syria experienced a severe drought in the run-up to the war, from 2006 to 2009. Over 800,000 people, most of them subsistence farmers, lost their livelihoods, driving internal migration to that country's urban centers, where most became unemployed. Within two years, civil unrest in those centers led to civil war.

Of course there are other explanations for these events and significant differences among countries. Experts disagree on the main causes. For the problem of tropical disease expansion, for example, the rise of airline travel is certainly a factor. For the problem of mass migration, economists tend to blame lack of economic opportunity or a poor business climate. Social scientists blame religious conservatism and lack of gender equality. Political scientists focus on poor governance and political instability. Syria's current civil war is closely tied to the "Arab Spring" that swept through the region between 2010 and 2012, and a population disenchanted with a dictatorial government. Hydrologists will blame falling water tables and poor water management. Demographers (people who study population growth) point to overpopulation, high birth rates, and lack of educational opportunities for women. Military and security experts tend to blame civil war, extremism, or the existence of criminal gangs that engage in human trafficking. All of these factors are relevant but (with the exception of demography) are not really underlying causes of mass migration. After all, there wouldn't be much profit in human trafficking if no one wanted to leave.

What's the link between climate change and mass migration? Consider African countries, such as Chad, Niger, Mali, and Mauritania, that are suffering water shortages and desertification. Rising temperatures and drought, together with rising population, have led to falling water tables, crop failures, and loss of income for rural families. Most of these countries lack a strong industrial base or major natural resources; agriculture anchors their economy. If you are a subsistence farmer here and lose your agricultural income, you can't walk away from your farm and land a factory job. The region is already sweltering for much of the year, and lacks reliable electricity for air conditioning. So the best and brightest leave, hampering development of a modern industrial economy. The resulting poverty fuels political instability, civil unrest, and a host of other social ills. Young people with no job and no prospects for a better life make easy recruiting targets for criminal gangs or extremist groups. It's possible for a country to overcome these challenges with good governance, infrastructure investment, and economic growth, but it is difficult.

Reports on the economic impact of climate change usually state that it will cause large future economic losses. These reports don't have much impact – most people have trouble focusing on nonspecific future events (recall our "present bias," mentioned in Chapter 1). Let's look at some losses that are already occurring. Since 1972, Whiskey Creek Shellfish Hatchery in Oregon has produced oyster larvae for the commercial oyster industry. Prior to

2008, it produced 7–10 billion oyster larvae per year. In 2008, production fell to 2.5 billion larvae, devastating the company and nearby oyster farms that depended on Whiskey Creek's production. Ocean water pumped into the facility had become too acidic for the oyster larvae because of rising atmospheric $CO_2$. Adding more alkaline water now costs the company $1 million a year. If Whiskey Creek's problems were caused by a nearby industrial source, they could sue to recover monetary damages. But for companies and individuals hurt by $CO_2$ pollution, it's impossible to identify a specific culprit. Existing laws and regulations don't cover cases like Whiskey Creek. For $CO_2$, the current rule is "Polluters don't pay."

I'll be the first to admit that I wouldn't want to be slapped with a big fine or tax to cover my carbon emissions. But eventually we'll all have to pay, in one form or another.

The 2012 heat wave and drought in the central US broke numerous records. In municipalities across nearly half of the lower 48 states, thousands of high temperature records were broken. The US Department of Agriculture declared parts of 26 states to be disaster areas, its largest ever designation, mainly because of the effects of drought on crops. Large parts of Colorado were also affected by wildfires.

It may have been difficult in the past to directly attribute a given heat wave to global warming, but it is now clear that the number and intensity of these events is increasing, in agreement with the predictions of most climate models. Given the intensity of the 2012 events in the US, it would take an especially strong "head in the sand" personality to deny the obvious link. Jonathon Overpeck, a scientist at the University of Arizona, stated it clearly to the Associated Press: "This is what global warming looks like at the regional or personal level."

In early July 2012, during the height of the US heat wave, I was driving across western Ohio, a major agricultural area in the US Midwest. At 8:30 p.m. local time, when one normally expects the evening heat to be cooling off, it was still 40°C (104°F).

The July 7 issue of the *Sydney Daily News*, a local newspaper, had five articles directly or indirectly related to the unprecedented heat, the drought, and a freak storm (a "derecho") that had raced across the central US a week earlier, killing at least a dozen people. No mention was made of possible links to global warming, perhaps reflective of the ambivalence many people in the US heartland feel about the topic. Below are the headlines, with parts of the lead sentence or sub-heading in italics:

1. Heat wave kills three: *Ohio's heat wave has turned deadly, a coroner said Friday, as he blamed excessive temperatures for the deaths of three people ...*
2. Excessive heat causes poultry deaths: *Excessive heat is being blamed for hundreds of thousands of turkey and chicken deaths ...*
3. Farmers grow concerned over dry conditions: *Concerns are growing among Ohio's farmers as abnormally dry conditions and triple digit temperatures scorch already parched fields, stunting much of the corn and soybean crop.*
4. Food replacement assistance available: *The Ohio Department of Job and Family Services has been granted a federal waiver to issue mass food assistance ... to 34 Ohio counties hardest hit by the June 29 storm and power outage.*
5. Country fried music: *Spectators brave one of hottest Country Concerts ever.*

Headline #1 draws a clear link between heat and fatalities, while headlines #2 and #3 indicate a link between excess heat and financial losses. Crop losses from this event across the US are now estimated at more than $8 billion. If it is true that global warming can now be linked to an increasing number of fatalities and large financial losses, a number of ethical, financial, legal, and political issues are raised, issues for which society does not currently have good solutions.

## SOME LESSONS FROM GREENLAND

If you visit the village of Illulisat on the west coast of Greenland, you will be treated to a spectacular display of icebergs. There are so many icebergs that even in the summer, its nearby fjord, which

FIGURE 8.9 The edge of the iceberg mélange near Illulisat, Greenland, probably marking the underwater location of an end-moraine, a ridge of gravel and rock representing the glaciers' farthest advance thousands of years ago.

measures roughly ~10 km wide by 100 km long, is choked with ice. It helps that the end of the fjord is blocked by a moraine, a pile of rocks and gravel emplaced thousands of year ago by the glacier when it was much farther out in the fjord than it is today (Figure 8.9). The scenery is so beautiful that this location was named a World Heritage site in 2004.

Several hotels in town provide balconies to sit and watch the iceberg spectacle. If you look at the ice for just a few minutes, you might be convinced that it is not moving. Perhaps everything is frozen or the icebergs are stuck on the bottom of the fjord (recall that most of the iceberg is submerged – that's one reason the Titanic got in trouble). But if you look few hours later, the scenery is subtly different. Time-lapse photography reveals that most of the icebergs are engaged in a beautiful ballet, responding to winds, tides, and currents in a complex dance. Occasionally, one will flip over, perhaps after melting and changing shape by some critical amount, causing large waves that can actually be dangerous to nearby boats and bystanders. Most of the time, however, the motions are subtle, not picked up at first glance, and only visible through the compression of time afforded by the time-lapse technique.

Illulisat's icebergs are coming from a glacier that is nearly 100 km away, Jakobshavn Isbrae. Jakobshavn is Greenland's largest and fastest moving outlet glacier, moving up to 50 meters per day. Its calving front (Figure 8.10) periodically produces spectacular displays

FIGURE 8.10A The calving front at Jakobshavn Isbrae during summer 2012. An instrument for measuring ice velocity is visible in the foreground, along with a small tent to shelter the operators (the author and graduate student Denis Voytenko) from the wind. The curved ice front is approximately 150 meters high, and its center is 5 km away, near the upper right hand side of this photo. The Greenland ice sheet is in the background and to the right of the ice front. Jakobshavn's ice-filled fjord is at the center-left, below the front. Ice motion is from right to left. Photo by the author.

FIGURE 8.10B A close-up of the calving front, taken near local midnight. The boundary (calving front) between the higher elevation glacier (right hand side) and lower elevation ice-choked fjord (left hand side) is more apparent when the sun is low on the horizon.

as huge icebergs cleave off the edge of the glacier and overturn in the fjord. The calving front has retreated many kilometers farther from Illulisat just in the last few decades, probably in response to global warming, but it still keeps the fjord filled with ice. Time-lapse photography is used in scientific studies to record glacier motions, calving events, the retreat of the ice front, and ice motions in the fjord (Figure 8.11). It was also used extensively in the beautiful 2012 documentary film *Chasing Ice,* which was produced by photographer James Balog.

On a year-to-year basis, changes in the planet's climate are imperceptible, just like the motions of Illulisat's icebergs are imperceptible if you look for just a few moments. Even over a lifetime, an individual would have trouble perceiving the warming of Earth's climate that almost all scientists are certain is now going on.

FIGURE 8.11 Scientists from the University of Alaska work on a time-lapse camera that records motion of the ice mélange in Jakobshavn's ice-filled fjord (background). At this location, the fjord is about 8 km wide. Photo by the author.

Evolution has not hardwired us to be sensitive to changes in our environment that operate so slowly. Plots like Figures 8.3, 8.4, or 8.5 are a bit like time-lapse photography – they let us observe longer time scale trends that we could not otherwise perceive. Viewed over these longer time scales, the upward trend of temperature is clear.

Earth has experienced natural warming events in the past, at least half a dozen in the last few million years. Even in the past 12,000 years, there have been periods almost as warm as the present. The big difference now is that the *rate* of warming is much faster than earlier, natural warming events. Going back to Figure 8.4, we can see some temperature spikes in addition to the longer term warming trend. For example, warming between 1970 and 2000 was about 0.75°C, equivalent to a rate of about 2.5°C per hundred years if it continued.

This is the crux of the problem from the standpoint of ecosystems and species survival. It is a subtle point, not emphasized enough in most climate change discussions. The pace of past warming events was slow enough that most of the planet's ecosystems and species could adapt. This time it's different. The rules of survival are not only changing, they are changing fast. Changes in average temperature, and more importantly, changes in the timing of change (seasons) are happening so fast that many species can no longer adapt. Hummingbirds migrating north may find that they missed a favorite spring flower and its nectar because it bloomed two weeks earlier. Hummingbirds need to eat frequently, so even one missed opportunity can be disastrous. If it just happened in one year, individuals would die, but the species as a whole would likely survive. But what happens if the species misses three or four spring blooms in a row? What happens if you combine this stress with habitat loss (the field of spring flowers was paved over to become a shopping mall) and other environmental stressors, such as impacts from herbicides (fewer flowers to eat) and pesticides (high levels in the environment can affect a bird's immune system)? Understanding species survival is a bit like interpreting the low elevations and flooding in New Orleans – there are a variety of geological processes that contribute to those low elevations, and they all add up. By the same token, all the various stresses on an individual animal add up to determine whether that individual survives and whether the species survives. Global warming is becoming a big stressor for a lot of species.

One issue that continues to dog the climate change debate is how much modern, human-induced warming differs from earlier, natural warming events? If we focus on temperature alone, the issue is not so clear. It has been warmer in the past. If instead we focus on perturbations to temperature, it's easier to see the human influence. Figure 8.12 shows how fast temperature has changed over the last 11,000 years. The present sticks out like a sore thumb. In round numbers, the natural background variation, at least for the last 11,000 years, is less than a few tenths of a degree celsius every hundred

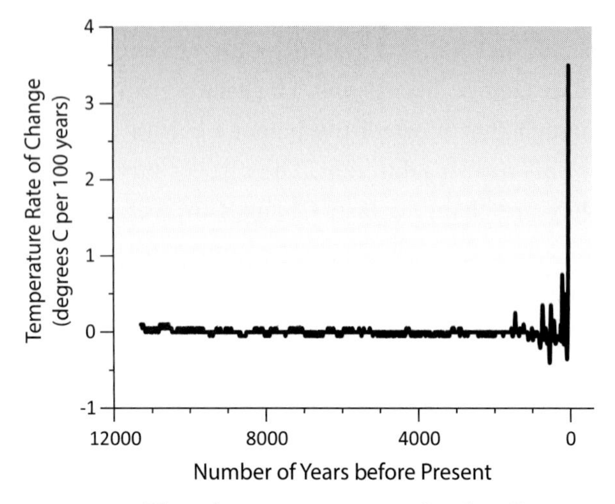


FIGURE 8.12 Plots of temperature versus time (e.g. Figures 8.4 and 8.5) don't tell the whole story when it comes to climate change and the planet's ecosystems. It has been much warmer in the past, unrelated to human activity. What is unique about modern warming is how *fast* it is happening – much faster than previous warming episodes, and too fast for many species to adapt. This figure shows how fast global temperature is changing (vertical axis) versus time. Note the very low rates of change until the last ~1,000 years. The apparent variability beginning about then may simply reflect more detailed information in the more recent period, but could also reflect the beginnings of human influence on climate, when deforestation and agriculture were becoming widespread (Ruddiman, 2013). More importantly, the modern era (last ~150 years) is clearly anomalous. Based on data from Marcott et al. (2013), calculated by taking the first derivative of their global temperature versus time data set.

years – that's the rate of change that most plants and animals are comfortable with, and can adapt to. In contrast, modern warming is ten times faster. This is too fast for many species, and they will go extinct. Humans living in some low-elevation equatorial regions, where it is already pretty hot and humid, close to our natural "survival window," may also have trouble adapting without access to air conditioning and other resources.

It's worth remembering that rapid environmental change played a big role in some of the planet's past great extinctions. The Cretaceous event that wiped out the dinosaurs, likely caused by

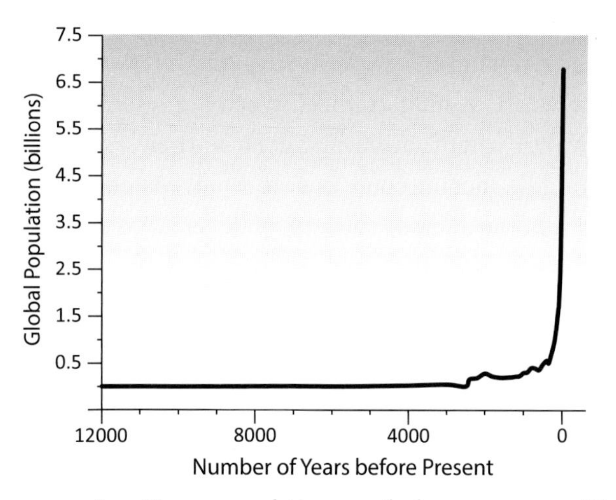


FIGURE 8.13 Human population over the last 12,000 years. The brief "dip" in population growth around 400–800 AD (1,200–1,600 years before present) and uneven growth in the next few centuries may represent the influence of the Justinian and later plagues, when Black Death (possibly bubonic plague) first began to devastate Europe and Asia. Data sources: United Nations; US Census Bureau.

a giant meteorite (Box 2.5) is a great example. It would be ironic if the next major extinction associated with rapid environmental change was caused by our own actions.

The plot of temperature change versus time (Figure 8.12) looks a lot like a plot of global population through time (Figure 8.13). This should come as no surprise. While correlation does not prove causality, the simplest explanation for the rapid temperature rise in the last few hundred years is the rapid rise in human population in the last few hundred years, combined with the rapid rise in our $CO_2$ "footprint" from industrialization. Some skeptics admit that the Earth is warming, but they don't believe humans are the main cause. For me, it's hard to look at plots like 8.12 and 8.13 and not see the connection.

The role of rapid population growth in our environmental problems and its solution (birth control and family planning) are sensitive topics in many cultures and religions. It is sometimes said that climate change is just a polite way of talking about overpopulation. This is an oversimplification, since a relatively small number of countries (the wealthy, industrialized West) are responsible for most of the $CO_2$ currently in the atmosphere. Nevertheless, a large number of currently middle-income countries are rapidly catching up with the industrial West in the dubious category of largest $CO_2$ emitters. These issues need to be openly discussed and thoughtfully addressed by all the world's nations if we are to avoid planetary catastrophe.

Many Earth processes are like Jakobshavn's icebergs. On short time scales, no change is apparent – but over time, a pattern emerges. Isaak Cline thought Galveston was safe from hurricanes because the city hadn't experienced one in 50 years. Managers of New Orleans' levees couldn't imagine their degraded flood defenses were too low to defend the city. The earthquakes discussed in Chapters 3, 4, and 5 may only show patterns when considered over thousands of years. Those who designed and operated the Fukushima nuclear plant got in trouble because they couldn't conceive of dangers on that time scale. Climate change skeptics say they can't imagine a change they can't see or feel.

We all need to start paying attention to the signals our planet is sending us. They are recorded by the "time-lapse photography" of science, whether it's average temperature, Arctic sea ice loss, ocean acidity, or species extinction. If we do start paying attention, we may be able to leave the Earth in a better state for our grandchildren – and theirs.

# 9 Solutions

The problems discussed in this book have common sense solutions, many of them presented in earlier chapters, and summarized at the end of this chapter. For readers who are pessimistic about the state of our world and our various challenges, it's important to recognize that, most of the time, things work out pretty well. Improvements happen slowly, but they do happen. The next building or piece of major infrastructure is usually better than the last one due to a combination of innovation and better regulation. The media tends to focus on disasters, which means we don't always hear about positive developments. For example, Space-X was the first private company to develop unmanned rockets to supply the International Space Station (ISS). During the launches, the company conducted engineering tests of soft landings for the first stage of its rocket booster, to allow for its eventual reuse. A January 2015 report in the *Washington Post* about the successful launch of a Space-X rocket to resupply the ISS had the following headline: "Space-X unable to land rocket at sea." Lyndon Johnson, US president from 1963 to 1969, once said "If one morning I walked on top of the water across the Potomac River, the headline ... would read: 'President Johnson can't swim'." It's important to recognize successes as well as failures.

When things go wrong, we can learn from failures and improve. Since the industrial revolution, society has made huge improvements in areas related to human welfare and scientific understanding. Such progress will likely continue. In this closing chapter I summarize our hazard-related problems, and describe some recent advances and common sense things we can do to speed up the improvement process and make life better for our grandchildren.

## COMMUNICATION

### *Improving Scientific Communication*

Scientists excel at ferreting out the mysteries of nature from sparse and incomplete data. But most of us are not very skilled at explaining our work to non-scientists, why it's important, or why a certain policy should be followed. There are exceptions. Carl Sagan, an astronomer from Cornell University, captured the imaginations of millions with his eloquent descriptions of the universe. Astrophysicist Neil deGrasse Tyson is following in his footsteps, tweeting humorous and thoughtful messages with a scientific flavor (one tweet from 2011: "If aliens did visit us, I'd be embarrassed to tell them we still dig fossil fuels from the ground as a source of energy").

One problem is that when asked to make policy statements, many scientists tend to speak to their peers, rather than the public. We use technical jargon, emphasize measurement uncertainty, and place caveats on our conclusions. This reflects a scientific culture that promotes cautious understatement over flamboyance. Common sense statements that simplify the problem are usually avoided. Policy statements from committees are even worse, often vague and convoluted because consensus trumps clear writing.

Scientists even have trouble speaking to other scientists who are not in their discipline. Most of our specialties are now so narrow that it's difficult for scientists not directly engaged in the same line of research to understand each other – there is simply too much technical jargon. Many scientific journals suggest to authors that they produce an abstract (summary) of their article that is understandable to non-specialists, but this advice is often ignored. The prestigious journals *Science* and *Nature* have recognized the problem, and now produce a one or two page summary of some of the significant papers on a given issue, written for non-specialists. The summaries are published in the "Perspectives" section in *Science*, and in the "News and Views" section in *Nature*. But another indicator that we have a problem in scientific communication is that *Science* now also produces an even shorter, one paragraph "summary of the summaries." You know we have a communications problem when even a summary needs a second round of simplification for readers, most of whom are professional scientists.

In fairness, some of this reflects a cultural shift to shorter news stories, communication bursts and tweets (messages of less than 140 characters). The ever-expanding rate of scientific publication also challenges any scientist who tries to keep up with developments even slightly outside his or her field. The shorter summaries are much appreciated, and some weeks, that's all I have time to read.

Many countries are pushing for increased STEM (science, technology, engineering, math) education in colleges and universities, as a way of promoting economic growth. But STEM graduates also need grounding in the liberal arts, including writing and speaking skills. It doesn't do much good if you've discovered the cure for cancer or the origin of the universe but can't explain it to anyone. Too many scientific papers are poorly written, violating basic rules of composition, with long tortured sentences full of acronyms and parenthetical statements.

Other factors make it difficult to get the message across. Sometimes scientists make mistakes, affecting the credibility of the larger enterprise. A recent report on global warming included an incorrect estimate for the year when Himalayan glaciers will disappear. The resulting media furor gave ammunition to critics of climate change science. It's easy to ignore advice if you think its wrong, especially if accepting that advice is going to cost you money. While mistakes happen in science as in any other human endeavor, incidents that seriously erode scientific credibility are actually quite rare. This reflects a built-in corrective aspect of science, where publications are peer-reviewed, and ideas are vigorously debated at professional meetings and in the scientific literature. Bad ideas usually don't last very long – scientists are a skeptical bunch.

Fortunately, scientists are getting better at communication. Professional meetings now offer seminars that help scientists improve

FIGURE 9.1 A beer glass commemorating a "Pint of Science" event, where well-known scientists give free lectures at a local pub. Photo by J. E. Dixon.

their communications skills with the press and public. In 2012, two scientists at Imperial College London, Drs. Michael Motskin and Praveen Paul, started an outreach event called "Pint of Science" (Figure 9.1). Lovers of beer and science meet at a local pub to hear an eminent scientist describe his or her research in non-specialist terms. The event now runs in nine countries. I recently attended one of the events in my hometown of St. Petersburg, Florida, and was amazed at the high turnout, enthusiastic audience, and range of interesting questions.

Communication styles that recognize human strengths and weaknesses can also help. Although we may be insensitive to long-term processes, evolution has also given us strong visual sensitivity, presumably related to our ability to perceive short-term threats and opportunities in the immediate environment, and determine the correct fight or flight response. We are good at pattern recognition: It's a shortcut that enables rapid evaluation of threats, allowing our brain to avoid complex calculations in an emergency. For this reason, pictures are far more persuasive than a table of numbers.

A report on the scientific background to the New Orleans disaster, published in 2007, was aimed at promoting the construction of more resilient infrastructure in the region. The report consisted of 29 pages of text. The only image in the report was a picture on the cover, consisting of a black and white satellite photo of the clouds from Hurricane Katrina. Perhaps it is not surprising that the report did not have a significant impact on the rebuilding effort or on people's thinking about long-term development in New Orleans (Chapter 7). Scientific committees need to recognize that their reports would have more impact if it were possible for people to read them without falling asleep. My suggestion for committees is to hand their "final" version of the report to communication professionals who can make it readable and add punchier graphics. A short executive summary is also a good idea. Anything longer than half a page is a good indication of a report written by a committee.

Humans are very good at recognizing *changes* in patterns, especially if they happen quickly. Again, this is probably related to the evolutionary advantage bestowed on those who can perform rapid threat assessment. In developing countries, one of the first things people buy when they reach a certain level of disposable income is a television. One reason why YouTube is so successful is our fascination with movies and animations. In 2012, one of the hot new apps for smart phones allowed people to download and share videos. Video sharing over the Internet is now common. Communication that makes use of animation is bound to be more successful than a presentation full of equations, tables, and bullet points on PowerPoint slides. A simple video that speeds up a slow, otherwise imperceptible process can be a valuable communication tool. Time-lapse photography and low-cost software for video production are now ubiquitous and are making scientific presentations more accessible to the public. At a recent professional meeting I attended, organizers gave prizes for the best videos; students invariably made the best ones. The worst was by a well-known professor whose video included a slow pan down several pages of equations.

I usually attend the annual meeting of the American Geophysical Union, a major scientific society. A decade ago, videos were rare at the meeting. By the time of the 2014 annual meeting, one out of every four or five oral presentations that I attended included an animation or short video, definitely a step in the right direction.

### *Improving the Listening Skills of Politicians and Policy Makers*

Scientists who are poor communicators are only part of the problem. Some politicians, some members of the media, and some members of the public also bear some responsibility for not listening to the message of natural hazard preparedness and other important scientific issues. In this section, I'll discuss a few examples from the political sphere.

Many countries are at risk of natural disasters in one form or another, and disaster preparedness should be an accepted part of the national character. Adroit assessment and management of risk and disaster should be seen as a political plus. Failure can be damaging to political careers. The Democratic Party of Japan, in power at the time of the Fukushima nuclear meltdown, received a drubbing at the polls the following year, in part due its perceived inability to deal with the crisis. In the US, the presidency of George W. Bush was hurt by his handling of Hurricane Katrina and its aftermath in New Orleans in 2005.

Since the United States experiences a range of natural hazards, you would think that its politicians would be especially sensitive to the charge of being "soft on natural hazards," especially after the Katrina example. In 2009, Bobby Jindal, governor of the state of Louisiana, questioned why "something called 'volcano monitoring'" was included in the economic stimulus bill that President Obama had signed earlier that year. Jindal said, "Instead of monitoring volcanoes, what Congress should be monitoring is the eruption of spending in Washington." Jindal's attitude is unfortunately typical of many politicians, who for the most are not very knowledgeable about natural hazards, the great harm they can do, and the way that hazards can be minimized through research, planning, monitoring, and infrastructure investment at relatively low cost compared to the subsequent expensive but potentially avoidable disaster.

Volcanoes are one type of hazard where low-cost monitoring can pay big dividends. Although we can't quite predict volcanic eruptions, we are getting pretty close. Current technology and understanding can give advance warning for most major eruptions, often with a lead-time of several weeks or longer. The 1980 eruption of Mount Saint Helens (Figure 9.2) in the US state of Washington is a good example. This was a major eruption, the costliest in US history, but advance warning by scientists allowed most people to get out of harm's way. Casualties were limited to 57 people (mainly people who had been warned but refused to evacuate), relatively small given the scale of the eruption. The 1991 eruption of Mt. Pinatubo in the Philippines similarly occurred with lots of advance warning, in part due to the availability of monitoring data.

A number of countries are located in or near subduction zones and are at risk from active volcanoes, including Japan, Indonesia, New Zealand, Chile, and the US. These volcanoes are capable of causing great loss of life and property, and economic disruption to the tune of billions of dollars. However, with proper monitoring, such losses can be greatly reduced. Volcano monitoring is also relatively inexpensive, hence cost-effective. It's hard to think of a better long-term investment than monitoring volcanoes that are at risk of catastrophic eruption and have the potential for generating significant fatalities and economic disruption, either because the volcano is close to a population center, or has the potential to disrupt air traffic (volcanic ash erupted high into the atmosphere is difficult for pilots to see, and

FIGURE 9.2 The May 1980 eruption of Mount St Helens. Photo courtesy of US Geological Survey.

can choke jet engines). Perhaps we should rephrase Jindal's statement: What the public should monitor are the eruptions of ill-informed politicians.

### *The Role of Media*

The media has a critical role to play in communicating scientific findings and hazards to the public, and communicating the achievements of politicians in terms of preparedness and response, or their failures. Unfortunately, few newspaper or television reporters have a scientific or engineering background, so coverage of such stories can be simplistic, naïve, or sometimes just plain wrong. Political reporters pride themselves on being able to catch politicians off guard with a juicy quote; they seldom return to that source for verification, fearing retraction or qualification. The same approach does not make for good science journalism, where a more collaborative approach between source and writer is beneficial. Accuracy on a complex topic may require several back-and-forth sessions between the reporter and the scientist to refine the story and get the science right. This can be challenging in the rapid response world of TV and newspapers, but social media are making it easier to get the story out. I often refer reporters to professional web sites where up-to-date, authoritative reports on the latest natural disaster and professional graphics are available. For example, earthquake information is available at the web sites of IRIS (www.iris.edu), the US Geological Survey (USGS, www.usgs.gov), the Seismological Society of America (www.seismosoc.org), the British Geological Survey (www.bgs.ac.uk), and the French Geological Survey (www.brgm.fr). The web sites of UNAVCO (www.unavco.org) and the USGS often post graphics and animations of the displacement fields and shaking associated with major earthquakes as measured by GPS and seismographs. Most people are amazed to learn how much the Earth actually moves during an earthquake.

The media are especially important when new scientific information exposes a significant hazard to society, whether it's leaded gasoline, tobacco, chlorofluorocarbons, mercury, air pollution, $CO_2$ emission, or poorly designed or located nuclear power plants. However, these hazards may be closely tied to profits for a certain industry, and special interests can be aggressive at defending their source of wealth. It can be difficult for scientists to get the message out in the face of determined opposition from these special interests. Not all journalists are skilled at separating fact from froth, and can be misled by these special interests, as discussed in earlier chapters. We owe a great debt to those reporters who persevere against these odds, sometimes at great personal cost.

### *Whistleblower Protection for Scientists*

In the financial world, individuals who come forward with evidence of tax evasion or criminal wrong doing can be afforded "whistleblower" protection by the government. This can include financial incentives as well as compensation for losing their job as a result of their disclosure actions. In the academic world, scientists usually have some protection against being unjustly fired (tenure, typically awarded to professors after five or six years of probationary employment). But in our increasingly complex society, there are many other ways that scientists who disclose uncomfortable truths can be harassed. Excessive demands for backup documentation or email, untruthful accusations of ethical violations, withdrawal of funding, threats of detailed tax audits, and even veiled threats to the personal safety of family members have all occurred.

It is hard for environmental and climate scientists to get the message out if they feel threatened by special interests. Right now it is too easy for special interests, or the politicians and private organizations they fund, to harass scientists who are honestly trying to alert the public to environmental dangers. I would like to see specific legal protections for scientists engaged in this type of research. A scientist who felt threatened would have a legal right to sue the harasser, and force disclosure of the financial backers of that harassment. Public prosecutors would have the authority to pursue criminal charges against individuals and companies supporting this type of activity. Because evidence of such activity is notoriously difficult to uncover (special interests are quite skilled at covering their tracks with offshore shell companies and supposedly independent "think tanks") the statute of limitations for such offenses should be lengthy.

In a similar vein, I would like to see political Ethics committees and the media do a better job of condemning politicians who use the powers of their office to discredit and harass scientists or the organizations who support their work. This harassment is often done at the behest of special interests who may have donated funds to that politician's re-election campaign. While this may be considered "business as usual" in many jurisdictions, to me it is a clear violation of ethics and taxpayer trust.

## TRANSPARENCY

It is often said that sunshine is a great disinfectant. Transparency, which allows light to be shed on a given issue, is closely related to communication – it's hard to have an informed conversation without knowing all the facts. More transparency could go a long way toward solving at least some of the problems identified in this book. I'll cite three examples, related to building safety, media reporting, and environmental audits.

### *Building Safety*

Would you rent an apartment if you knew that it was not engineered to withstand levels of earthquake shaking that had a 10 percent chance of happening sometime in the next 25 years? You might, if the rent was cheap, or if you were single, or if you only needed it for a few months. On the other hand, if you had a young family, and needed a place to settle down for the long term, you might think twice. A business owner looking for office space might go through a similar evaluation process. A small start-up company might opt to take the risk if the rent was cheap and money was tight; larger businesses with numerous employees looking for a stable, long-term location would probably look elsewhere. The point is, in both cases the parties involved could evaluate the situation for themselves if they were armed with the necessary information. Unfortunately, it is all too common for people to be unaware of critical information on building safety. Very few of those killed in the 2009 L'Aquila earthquake were aware that the buildings they lived in were known to be unsafe even in a moderate size earthquake. Although such a report had been prepared by scientists and engineers, municipal officials and local media did not publicize the findings, perhaps fearing economic or political repercussions.

It should not be difficult for every municipality in an earthquake zone to require building owners to notify tenants of the safety status of their building, as assessed by qualified engineers or building inspectors. The city of Seattle's Department of Planning and Development posts maps online that are regularly updated (www.seattle.gov/dpd) showing the locations of unreinforced masonry buildings, the kind most at risk from earthquake shaking. Brian Tucker, founder of the non-profit Geohazards International, has suggested special plaques that would confirm a building's earthquake resilience. With this kind of transparency, renters could choose the level of risk they were willing to endure, and building owners would have a financial incentive to upgrade their buildings. Those that did would be able to charge more rent, while those that failed might have trouble finding tenants.

### *Media Reporting and Political Statements*

In 2010, I published a scientific paper in the journal *Nature Geoscience* on the accelerating uplift of Greenland (due to loss of ice) with then graduate student Yan Jiang and colleague Dr. Shimon Wdowinski. At the end of the paper, we put the following statement in the Acknowledgments: "This work was supported by grants from ONR, NSF and NASA. Y. J. was supported by a NASA Fellowship." There was an additional section where we could list other information on potential financial or other conflicts of interest. We stated, "The authors declare no competing financial interests." Medical research published in *Nature* or its affiliated journals sometimes have one or more authors who received support from the pharmaceutical industry. These sections give the authors a chance to lay their cards on the table, so that readers can be aware of potential conflicts of interest, and evaluate the data and the authors' conclusions accordingly. The practice of acknowledging financial support in research publications has long been standard; additional statements on other potential conflicts of interest are also becoming common. It does not necessarily stop all abuses, but it goes a long way. Any author who knowingly and falsely declared no financial conflicts, and was later caught, would suffer severe damage to his or her reputation, would likely be barred from receiving future research funds, and might even be fired. Research ethics are taken seriously at reputable academic institutions.

I think the same standards should apply in journalism and politics. Some people will respond that this is not realistic, but that implies that we have already become fatally cynical about the state of mass media. In other words, we expect most items in the newspaper, television or Internet to misrepresent facts, slanted by the writer or speaker to reflect the views of whoever is paying the bills. Surely we can do better.

Here's an example of how such a system could work in journalism. A journalist interviewing a scientist for a story on global warming or some other environmental issue should find out the qualifications of the scientist conducting the research, and ask how the research was funded. If the journalist subsequently seeks an additional or alternate viewpoint on the story, the qualifications of the person expressing that viewpoint should also be stated, as well as the source of financial support for that person, whether it's an individual, a company, or an industry lobbying group. To say that support was provided by the Committee to Save Babies or the Institute for Sound Government is not sufficient detail; we need to know who specifically funded the project that the spokesperson worked on, the norm in scientific reporting. Failure to disclose should be considered an automatic red flag, a clear indicator that the interviewee has a conflict of interest that he or she is trying to hide. In such cases the interview should not be aired, since it is almost certainly biased and misleading. If for some reason the reporter has to air or publish the interview, it should be clearly stated at the beginning that the interviewee refused to disclose his or her funding source, and hence is probably trying to hide something. This code of transparency would go a long way toward ending the obfuscation and confusion that surrounds many current public policy disputes related to scientific and environmental issues, where a few narrow special interests tend to dominate the public discourse.

The same standards should apply to political speech. A politician espousing strong views on some scientific or environmental issue should be required to disclose relevant campaign contributions that may have influenced his or her opinion. Failure to disclose should result in the same sanctions that scientists face in scientific publication. If I had refused to name the sources of funding or declare my financial conflicts, *Nature* would have refused to publish my paper. By the same token, politicians who refuse to disclose relevant campaign contributions should not be able to use the airwaves, print or other media for their political speech. When politicians vote on a piece of legislation, a parallel declaration of any financial conflicts such as campaign contributions from groups with a vested interest in that legislation should be forwarded to the press as a routine matter. The media now spends lots of time ferreting out the sexual orientation and sexual peccadillos of politicians; why not spend some of that energy ferreting out financial conflicts. Let the public decide if the views and legislative votes being offered by our politicians are a rational solution to our problems, or simply reflect those of a special interest group with something to gain.

We don't need our politicians to pass laws requiring such disclosure. We just need journalists to ask the tough questions, and follow the money (see Chapter 6).

Unfortunately this is not always possible. I am painfully aware that there are a lot fewer journalists these days able to ask the tough questions. Since a well-informed public is key to a healthy society, this is a real problem, one for which I don't have a good solution. In non-democratic societies, journalists are often censored, imprisoned, or even killed for speaking out on sensitive issues or against special interests. In these situations, scientists may have to shoulder more of the responsibility for communication. Scientists outside the country in question have a special responsibility to speak out, as they can do so with much less personal risk.

The rise of internet-based media and consolidation of the newspaper industry into a few large for-profit companies has also greatly reduced the number of investigative reporters, and greatly reduced the ability of journalists to challenge special interests. My hometown newspaper, the *Tampa Bay Times*, represents a possible solution to the downsizing problem. The paper is owned by the Poynter Institute, a not-for-profit organization devoted to media and journalism studies, and a free press. The institute was founded by Nelson Poynter, the original publisher of the newspaper, who foresaw the need for independent media. Perhaps this model could be copied elsewhere. Alternately, we need a few civic-minded billionaires who would rather own a newspaper instead of a second yacht, and then agree not to interfere in the journalistic process.

### *Environmental Audits*

Publicly traded companies are supposed to have their financial activities periodically audited, with results made public, so that investors can make rational decisions about investing in the company. The auditors are supposed to be independent, but in several countries, including the US, the system is rigged: The companies pay the auditors for their work. This is an obvious conflict of interest, since the auditors want to keep their large corporate accounts. Not surprisingly, companies usually get positive reports on their financial health. Many of the firms that went bankrupt during the 2008 financial crisis had received glowing reports from their auditors the previous year. Enron, a large oil trading company that went spectacularly bust in 2001 after years of fraudulent activities, was consistently rated highly by its auditor, Arthur Anderson, one of several major global accounting firms. At the time, Enron was the largest corporate bankruptcy in the US, and one of the largest in the world.

If we can audit the financial books of large companies to protect investors, why can't we audit the safety and environmental practices of these companies to protect a much larger group of citizens? Safety and environmental audits should be conducted by independent entities, unlike financial audits, which, like the Enron example, are still generally paid for by the companies being audited. Imagine if a mandatory, independent safety audit had been conducted at TEPCO a few years prior to the Fukushima disaster. The company may have complained about the interference, but if it were required to act on the advice from such an audit, it could have saved itself and the nation of Japan billions of dollars.

## LET MARKETS WORK THEIR MAGIC

### *The Sea-Level Rise Example*

The basic principles of economics and free markets can solve some of our infrastructure problems if allowed to function properly, especially if we can improve information flow and transparency to help people make informed decisions. Let's look at the example of coastal flooding and sea-level rise to see how market economics might improve things.

Sea-level rise is one of the thornier problems our society is likely to face over the next 50 to 100 years, in part because of timing: Not much happens for a long time, inducing complacency, and then a big flood occurs, destroying lives and businesses and costing billions of dollars. This problem has implications related to health and safety, economics and business, legal issues related to the trade-offs between private property and public benefit, levels of public debt, treatment of refugees and even national security. Given high levels of debt in most western countries, it is unlikely that all cities at risk will be able to build New Orleans–style levees to hold back the water. But if we allow inundation of some areas, what happens to businesses or private landowners who stand to lose what, in many cases, could be their livelihoods and life savings? At the present time, flood events at a given location are infrequent enough that it usually makes sense to simply rebuild. But will this always be true? It seems reasonable to rebuild from a flood that only happens every few decades. But what if similar storms begin to strike the region every five or ten years? What if catastrophic flooding became a yearly occurrence?

Sometimes the best thing to do is simply to let the markets work. The insurance industry is pretty good at pricing risk, including the long-term risks associated with climate change and sea-level rise. Businesses and home owners can either choose to pay for the necessary flood protection, or decide its not worth it, sell out, and move to higher ground. The next buyer can then choose to either pay the higher premium, or make the necessary investments to change the business model or elevate the building. Although few people want to admit it, some low elevation areas will have to be allowed to revert to a natural state – it will be too expensive to save everything. In wealthy countries, governments could make this process less painful through tax incentives for people who donate their coastal properties to state or local parks or private groups such as Nature Conservancy. Poorer countries like Bangladesh may not have this option. However, agreements at the UN Conference on Climate Change in Paris in 2015 (COP 21) provide for financial aid for poorer countries to adapt to climate change. Perhaps some of this aid could be channeled into similar schemes. One thing to keep in mind during these difficult transitions is that coastal land that is "given up" is not necessarily lost to economic productivity – wetlands and shallow coastal areas represent important habitat for fisheries and other wildlife, and can be important future tourist attractions.

One barrier to fully functioning markets is lack of information, or sometimes even wrong information. In my own state of Florida, government employees are discouraged from using terms like "global warming" or "sea-level rise" in official communications. Another barrier is the presence of government subsidies that promote bad design and building practices in vulnerable coastal areas. These subsidies are both explicit and implicit. Explicit subsidies include subsidized flood insurance, and publicly funded sea walls, beach enhancements and other flood barriers. Implicit subsidies include the essentially guaranteed government help after a disaster that promotes rebuilding in the same area. This functions much like low-cost insurance: The cost of a disaster is difficult for a given community to cover, so the national government steps in, effectively spreading risk across a larger taxpayer base. This makes sense as long as the costs don't get too high, and the government can afford it. But as Western governments become increasingly indebted, as the number of flood events increase, and as the cost of each flood increases, something will have to give. One problem is that a subsidy, once offered, is politically difficult to withdraw. Imagine the outrage of citizens in a flooded area if their national government, after years of providing rebuilding aid to other disaster zones, suddenly decided to change the ground rules and not provide such aid.

If this system is to change, it has to be done gradually, so that individuals and communities can adapt and plan accordingly. What would a common sense flood insurance or disaster assistance program look like? One change might be that rebuilding aid comes with strings attached, including stronger requirements to upgrade infrastructure to higher standards of flood-resistance. In theory these standards already exist in the US, but they are often ignored or weakly enforced in the aftermath of a disaster. Alternately, insurance premiums could be lower for people or businesses that agreed in advance to rebuild in a safer location. If advertised well in advance, people could make appropriate economic choices. They would know that if a disaster struck, they would be rebuilding in a different spot, and factor such decisions into long-term plans. And there should be limits on rebuilding aid – perhaps the same building should only be eligible for one tranche of flood aid. An owner selling such a building would have to disclose to potential buyers that the property had already been damaged in a flood, had received aid, and was therefore not eligible for any further aid. In vulnerable areas, aid should only be allowed to cover relocation costs (up to some limit; we don't need public subsidy of waterfront mansions), and preclude rebuilding in the same place. This would allow some coastal property to revert to its natural state, increasing green space or public beaches. Publishing an official timeline for such gradually tightening standards and a "sell by" date would help individuals and businesses make appropriate investment decisions. Many locations along the US eastern seaboard, vulnerable in the long run, probably have many decades before the next flood, which is close to the economic lifetime for most of these buildings. If everyone knew with certainty that after the year 2050 (34 years from the time of writing this book) flood insurance would no longer be available for certain vulnerable parts of the US eastern seaboard, everyone could plan accordingly.

### *Tax Mercury and Gold*

Many countries put steep taxes on products that are deemed hazardous to human health, such as tobacco and alcohol. In places where marijuana has been legalized, it is also heavily taxed. These so-called "sin taxes" are meant to modify people's behavior for their own good. Why not apply the same principle to a range of toxic substances, for example mercury. As outlined in Chapter 6, mercury is extremely toxic, and is especially harmful to children and pregnant women. Anyone using mercury in an industrial process (such as battery makers or gold miners) would pay a special tax on purchases of this substance, providing financial incentive to look for alternate materials in the manufacturing process, or practice extreme recycling, where close to 100 percent is captured and reused. Likewise, any one emitting mercury into the environment as a by-product of an industrial process (such a burning coal for power) would pay a special mercury emissions tax, giving similar incentives to reduce emissions. Many countries signed the Minamata Convention in 2013, a legally binding treaty to reduce mercury emissions that will come into force once at least 50 countries have ratified it. However, the treaty does not specify exactly how countries should meet their treaty obligations. A tax on mercury would be a relatively painless way to do this, much easier to enforce than a host of regulations. It would generate revenues for cash-strapped governments, and unlike most taxes, should be an easy political sell ("Lets tax those who are poisoning pregnant women and unborn children").

After the 2008 global financial crisis, there was some discussion of a financial transactions tax, as a way to inhibit some of the more speculative trading that may have contributed to the crisis. Most economists criticized this proposal because it would inhibit investment in worthwhile economic activity such as job creation and infrastructure upgrades. However, not all investments contribute equally to societal welfare. A wealthy investor who puts a million dollars into a factory that produces medical devices is generating both jobs and useful products. If that same investor chooses instead to invest his or her capital in gold, the product of his investment likely sits in a vault gathering dust, generating few jobs, and no useful products. A significant fraction of gold that is mined today goes into passive investment products such as bullion or gold coins. The resulting additional demand inflates the price and encourages more mining activity.

The production of gold often involves the use of mercury in the refining process. So-called "artisanal" gold production recently surpassed coal as the major source of mercury released to the environment (Streets et al., 2011; Krabbenhoft and Sunderland, 2013; Lubick and Malakoff, 2013). Artisanal in this context refers to small-scale gold mining, which may be less regulated than large-scale operations. In the worst cases, mercury is simply poured onto a pile of rock and dirt and allowed to dissolve a fraction of the gold. The resulting slurry is burned to drive off the mercury and recover the gold, while the rock and dirt, now coated with mercury residue, is dumped as waste. Thus, most of the original mercury is dispersed into the environment, either into the atmosphere or local streams. Of course, not all gold comes from such environmentally destructive practices, but if you buy a gold coin it's hard to know where it came from. One solution to promote better refining techniques would be to generate a rigorous "paper trail." Producers of gold coins and jewelry with suitable documents could advertise their product as "mercury free" to the environmentally savvy, much as conflict-free diamonds are now marketed to politically correct young couples. A tax on gold transactions would also be useful, reducing demand for this metal and promoting more productive uses of capital.

### *Limits to Limited Liability?*

Limited liability is the concept that investors in a business are only liable for their initial investment if things go wrong and the company goes bankrupt. It is meant to encourage investment and create jobs and wealth. Economists credit this idea with the rapid rise of modern Western economies. The concept goes hand in hand with the idea that businesses should be allowed to fail if they are poorly run or if economic conditions change such that a business is no longer viable. The resulting bankruptcy allows the assets of these businesses to be redistributed, so that new businesses can spring up in their place and make better use of reallocated assets and labor.

After the 2008 global financial recession, it became apparent that some banks and other financial institutions had become so large that allowing them to go bankrupt would threaten the entire financial system; they were "too big to fail" and thus required government support, even if they had been poorly managed and otherwise deserved to go bankrupt. Since the 2008 financial crisis, some countries have required big banks to increase their financial reserves, in effect a financial penalty for being big and an attempt to reduce the risk of future failure.

I suggest that in our modern, industrialized economy, some businesses are now "too dangerous to fail." The products they produce, such as pesticides or other dangerous chemicals, or their by-products, may pose special hazards. Examples include pollution from an industrial process, or radiation from a nuclear power plant. TEPCO has already produced damage far in excess of its market capitalization, and hence is technically bankrupt, remaining in operation only because of taxpayer support. Imagine the damage that TEPCO could have inflicted on the Japanese population and economy if the wind, instead of blowing out to sea when its reactors exploded, had been blowing toward Tokyo?

On July 6, 2013, a 74-car freight train carrying crude oil derailed in the town of Lac Megantic, Quebec, in Canada. The resulting explosion destroyed the town and killed more than 40 people. The spilled oil will contaminate soil, local streams and groundwater far into the future. The cost of clean-up, reconstruction, and liability will likely be several hundred million dollars, and long-term environmental costs are unknown. These costs should be paid by the company, Montreal, Maine and Atlantic Railway (MM&A), which had profited from the transport of oil through the town. MMA declared bankruptcy in August 2013, allowing the owners to avoid paying for clean up, reconstruction and liability. These costs have now been shifted to taxpayers. Clearly, MM&A was too dangerous to fail, but fail it did. A depressingly similar incident happened in January 2014, when Freedom Industries, a US firm, spilled thousands of gallons of a toxic chemical used in coal processing into the river that supplies drinking water for Charleston, West Virginia. The company similarly declared bankruptcy less than two weeks after the spill to avoid paying clean-up costs, asserting that the accident was unavoidable because a frozen water line had inadvertently contributed to puncturing the storage tank. As Paul Barrett, a reporter for *Bloomberg*, stated "Shouldn't steel storage tanks containing dangerous chemicals be able to withstand the consequences of winter weather?"

What are some solutions for companies that are too dangerous to fail? Perhaps more transparency should be required, so that the public and media are able to examine their operations and safety procedures (see Environmental Audits above). In MM&A's case, that might have led to inquires about the safety of their tanker cars, the company's cost-cutting practice of having only a single engineer on large trains carrying dangerous material, or whether the brakes worked. In the case of Freedom Industries, poorly designed or maintained storage tanks likely contributed to the accident (such tanks should be double-sided). Independent safety and environmental audits with yearly updates should be required of such companies, much as independent financial audits should be required for publicly traded companies.

Perhaps such companies should also be required to post "Disaster Bonds," in effect an increased cost of doing business reflecting the increased environmental risk of their activities. The bond would be used in the event that the company declared bankruptcy, and should be sufficiently large to cover the cost of possible future environmental damage. The amount would be determined by an independent group, and for example could cover the cost of the maximum plausible accident, or some fraction of that value. If the company is not financially capable of posting such a bond, it should not be allowed to engage in dangerous activities. Some countries already require mining companies to post bonds covering the cost of future environmental clean up. Extending the concept to other industrial activities makes sense, although it would increase the cost of doing certain kinds of business.

## THE MAGIC OF NEW TECHNOLOGY: BLUE LEDS, THE TESLA, AND PHOTOSYNTHESIS

Research and new technology will help address many of the problems discussed in this book. For example, there has been considerable progress on making buildings and other infrastructure safer. "Smart" buildings, bridges, and dams are being developed that record internal strain (changes in shape) using sensitive strain cables embedded in the structures. These can help engineers determine when a structure is approaching failure. Research on earthquakes and volcanoes will improve our ability to forecast these devastating events. The 2011 earthquake and tsunami in Japan gave researchers a much clearer idea of the importance of marine geodesy, the ability to measure strain accumulation on the sea floor near ocean trenches (Newman, 2011; Dixon et al., 2014). In Chapter 6, I discussed new developments in the area of solar energy. In this section I focus on several new technologies that bear on the use, generation and storage of electrical energy, each of which has potential for reducing our $CO_2$ emissions and improving the planet's long-term health.

### *Blue LEDs*

The 2014 Nobel Prize in Physics was awarded to Drs. Shuji Nakamura, Isamu Akasaki, and Hiroshi Amano. Their citation reads: "For the invention of efficient blue light-emitting diodes which has enabled bright and energy-efficient white light." Incandescent light bulbs have been around for more than 100 years. A commercial design was patented by Thomas Edison in 1880. They are quite inefficient – most of the electrical energy goes into waste heat. Light-emitting diodes (LEDs) have been around since the 1960s, and in contrast to incandescent bulbs, are far more efficient. Unfortunately, until the work of Drs. Nakamura, Akasaki, and Amano in the 1990s, LEDs only came in two flavors (red and green) and it takes at least three (red, green and blue) to make white light, the kind that most people want in their houses. LEDs that give off white light are already available in stores (I recently purchased some). Within the next decade they will likely replace most incandescent and fluorescent lighting, promising significant decreases in overall energy consumption.

### *The Tesla*

Electric cars have also been around for a long time. They are much cleaner than gasoline or diesel in terms of air pollution. In theory they can also be cleaner in terms of their overall carbon footprint, depending on the source of electricity (very clean if the source is wind or solar; not so clean if it's coal).

A decade ago, if you wanted to buy an electric car, you had few options, all of them unpalatable. The main problem was range – electric cars simply did not go very far between charges. You had to be a real enviro-geek to pursue the electric vehicle option. Hybrid approaches like the Toyota Prius or Honda Insight, where a gasoline engine keeps the batteries charged, represent a partial solution to the problem, but lack elegance – basically you have to buy and maintain two engines. My family owns a Prius, sometimes ridiculed by our friends as the "Pious," i.e. a vehicle that allows the customer to feel good about his or her environmental contributions, but is not particularly efficient in terms of overall environmental impact.

This picture changed dramatically with the Tesla, a sleek, fast, all-electric car with a range of nearly 400 km (250 miles). The Tesla first went on sale in 2008. It is named after Nicola Tesla, a brilliant engineer from Serbia who immigrated to the US in 1884 and developed the concept of alternating current, used in most modern electrical grids. One of the innovations pioneered by the modern car company, Tesla Motors, is the battery, which has allowed the vehicle to break the range barrier, and along with it, most of the prejudices against electric cars. Owning a Tesla is now the "in" thing to do. Moreover, the company has decided to make a big investment in battery production to bring down costs. It is building a factory in Nevada with the goal of producing large numbers of the latest in battery technology. Improved battery technology and lower cost has implications far beyond electric cars – one of the major barriers to rooftop solar power is the inability to store energy when the sun is not shining. Cheap batteries will be a game changer for solar power, and may be the key to breaking our addiction to electric power derived from burning coal, oil and natural gas. Tesla recently announced that its batteries will be available for use in home storage.

### *Make Fuel Like a Plant*

Plants take energy from the sun, and use it to make various sorts of sugars, a type of carbohydrate they use for food. They use things like water ($H_2O$) and carbon dioxide ($CO_2$) as raw materials. Imagine if we could similarly make a fuel such as methane ($CH_4$) for our planes, trains, and automobiles using solar panels and the same common molecules:

$$2CO_2 + 4H_2O \rightarrow 2CH_4 + 4O_2$$

The main byproduct would be oxygen ($O_2$), the stuff we breathe. The reason this reaction doesn't happen outside the plant world is that it takes a lot of energy to split apart $CO_2$ and $H_2O$ molecules. Plants are good at it (the process is called photosynthesis), but humans have not yet mastered the trick, at least at low cost. In theory solar panels could provide the necessary energy. Many researchers are currently working on this problem, or the related problem of making hydrogen fuel with solar power. In the short run, these processes may prove to be an economical way of storing excess energy from wind and solar power installations, during periods when the wind is blowing strongly or the sun is shining brightly and the local electrical grid can't absorb the excess power. Hydrogen fuel cells are also being considered as a non-polluting power source for cars and trucks – their only waste product is water vapor.

### *Don't Give Up on Nuclear*

While research into new ways of generating energy is important, it would be wise to not give up on nuclear energy just yet. Since the Fukushima accident, nuclear energy has become even less popular with the public and with investors. But it represents an important source of energy with near zero carbon emissions, possibly a critical source over the next few decades as we struggle to come up with sustainable alternatives.

There may be better kinds of nuclear energy. India is researching a nuclear fuel cycle involving thorium. Thorium is more common than uranium (the fuel used now in nuclear plants), produces less high-level radioactive waste, and poses fewer problems in terms of proliferation of bomb-making material. Thorium-fueled nuclear reactors might be able to replace the dozens of coal-fired plants that India is otherwise likely to build in the next few decades as the country becomes wealthier and its citizens demand access to energy.

## R&D AND TAXES

The items listed above are just a few examples of new technology that can improve our planet's (and our own) long-term health by finding better ways to make and use energy. We need a lot more of them, and the best way to get them is through research and development (R&D). While companies can foot the bill for applied research and development of new products for the market place, it is up to governments to support basic research. This takes money, and a logical way to fund it is with a carbon tax. While such taxes have been levied in a few places, they are still pretty rare – no one likes taxes, and taxes on carbon have proven to be a tough sell in most countries. Perhaps if we used terms like "pollution tax" or a "tax to reduce childhood asthma," it might be more politically salable.

## AVOIDING THE "COAL TRAP"

The tropics and sub-tropics represent a disproportionate share of Earth's poorest nations, and will be hit disproportionately by the negative impacts of climate change. If wealthy northern countries hope to reduce the problem of mass migration from these regions, thoughtful development is required to bring these densely populated regions out of poverty. Hopefully this can be done without the environmental mistakes made by the current group of wealthy nations. Better management of water resources, effective development aid, improved educational opportunities for women, and access to low cost family planning would help.

Fundamental changes in the ways that energy is generated and used are also required. At the risk of oversimplifying, the world's wealthy nations became wealthy by first burning lots of coal to generate cheap energy, build infrastructure, and move up the development "ladder." Along the way, they poisoned a lot of their own citizens with air pollution, and set the world's climate system on a trajectory that looks increasingly bleak. The rich world needs to help poorer nations avoid the coal trap by enabling access to clean, low cost, low carbon energy sources. Direct subsidies are costly and inefficient. Sharing of intellectual property, or better yet, development of new intellectual property through joint development, are better bets. Because of their location, solar energy would seem to be an obvious energy source for many of these countries. Development aid focusing on local R & D in solar systems, exploiting the region's intense sunlight, would give these countries a financial incentive to pursue clean energy.

In the long run, Europe's current migrant crisis will only be solved through development that improves the lives of citizens in North Africa and the Middle East. Why not imagine (and invest in) North Africa as an energy powerhouse, not through continued pumping of oil, but through exploitation of that region's other energy resource, sunshine. If solar energy is feasible in cold, cloudy Germany, surely it is feasible in Tunisia, Morocco, Algeria and Libya, where the sunshine is more reliable. Skeptics will say that the distances are too large for economic export of electricity (transmission losses increase with distance). But research is leading to improved transmission efficiencies, and there are more ways to export energy besides sending electrons through a wire. Iceland exports its abundant geothermal and hydro-electric power via aluminum metal – smelting of this substance is energy-intensive, and is only economic close to cheap sources of electricity. As car manufacturers strive to increase fuel efficiency, aluminum will increasingly replace heavier steel as a material in automobile bodies. Europe's automobile industry could be a big beneficiary of such low cost aluminum.

Central America represents a similar opportunity in the western hemisphere. In addition to solar, that region has great potential for geothermal energy, as a result of its location on the Pacific "Ring of Fire" and the presence of numerous active volcanoes (Figure 2.3). El Salvador, one of the hemisphere's poorest nations and one of its smallest, hosts more than 20 active volcanoes. The region's abundant rainfall also contributes to the economic viability of geothermal power: Freshwater keeps the geothermal fluids dilute, limiting the concentration of dissolved solids and reducing mineral formation and fouling in the geothermal plant's heat exchangers.

The world's wealthy nations need to think about integrated solutions to the linked problems of clean energy, economic development aid and mass migration. Fostering energy R&D in places that need development aid and need clean energy solutions to grow their economies is one way to do this.

## INDEPENDENT BOARDS

John Hofmeister, a former president of Shell Oil Company, in his book "*Why We Hate the Oil Companies*" noted the following problem. Energy infrastructure projects typically take a long time to build, often one or two decades. But political cycles in most countries are short: At the US federal level, two years for a House member, no more than eight years for a president. The net effect of these different time scales is that it is often difficult to undertake the necessary planning and construction of new energy infrastructure.

All large projects generate controversy. Few politicians want to go out on a limb for something that, if successfully completed, will happen after his or her term of office is completed. Any politician who does otherwise reaps all the negative publicity at the beginning of the project, and none of the positive publicity at the end when the ribbon gets cut. Consequently, nothing happens for long periods of time, until a crisis hits.

Consider the example of old, inefficient coal-fired power plants that pump out high levels of pollution compared to modern, more efficient and cleaner-burning units. Regulations in many countries allow these older plants to continue to operate. In the US, air pollution laws have had the perverse effect of requiring that we keep older, inefficient polluting plants in operation, because plants only need to meet new standards once a cost threshold for upgrades is exceeded. The financial incentive is to do the minimum possible in terms of repairs and upgrades, an energy and environmental "own goal," giving us the worst of both worlds. The existing political process promotes this unsatisfactory status quo.

Hofmeister, noting the success of the independent Federal Reserve Board in the US ("the Fed") at regulating money supply and interest rates, proposes an analogous Federal Energy Resources Board ("the Ferb"). This board would be composed of experts in energy, economics and the environment, appointed for long fixed terms. It would take most energy-related decisions out of the hands of politicians, and would have the power to make the compromises and investments required to safeguard the nation's environment, economy and energy independence (yes, it's possible to have all three, at least if one is willing to make some compromises as well as the necessary long-term investments and tough short-term decisions). I could imagine such a board quickly closing down the "dirty dozen" of coal fired power plants, or at least requiring immediate pollution control upgrades. I can also imagine such a board approving Keystone, a controversial pipeline that would bring oil from the Canadian oil sands in Alberta to underutilized Gulf Coast refineries, but requiring the imposition of a carbon tax. The tax would better reflect the full environmental and health-related costs of burning fossil fuel. Environmentalists do not support Keystone, because oil derived from these deposits is typically higher in carbon emissions compared to "standard" oil. However, most newer sources of oil ("unconventional" oil) are more expensive to produce and have higher carbon footprints compared to deposits which in the past could be pumped out of the ground easily and at low cost. Without the pipeline, oil from Alberta's oil sands and much US unconventional oil is now brought to market by rail, which has a much higher carbon footprint compared to pipeline transport and is more dangerous. With a carbon tax in place, and the more realistic pricing structure it implies, markets could decide whether oil from Keystone or other unconventional oil deposits was justified economically and environmentally, using full cost-accounting principles.

I like Hofmeister's idea. I think it could be useful in many countries, and could extend to a number of other areas and infrastructure problems. We already have one working example in the US, the Nuclear

Regulatory Commission (NRC), which oversees the nuclear energy industry. After Fukushima, NRC mandated a review of nuclear power plant safety, including "lessons learned." It seems to be working – Hurricane Sandy in October 2012 flooded several nuclear plants, with no adverse affects. It has some limits (NRC has not been able to mandate opening of the Yucca Mountain repository for high level nuclear waste) but is generally credited with insuring safe infrastructure for the US nuclear industry.

## LESSONS LEARNED

Learning lessons from failures is an important aspect of improving infrastructure resilience. No system is perfect, and mistakes happen. But if we can learn from our mistakes, our children and grandchildren will have a smaller mess to clean up, and smaller bills to pay. Many organizations conduct "After Action" exercises to see what worked and what didn't. The US military started to conduct a broad review of what went wrong with existing infrastructure, pre-event planning and post-event response to Hurricane Katrina, but the effort was soon refocused to look just at levee failure. This is unfortunate – as the costliest disaster in US history, we might have learned some useful things from such an After Action exercise. A full and well-publicized study might also have prevented the Fukushima disaster, by highlighting the importance of planning for high water in various backup systems.

In the US, an agency known as the National Transportation Safety Board (NTSB) is charged with investigating all civil aviation accidents, and making recommendations to pilots, airlines, and airplane manufacturers to improve safety. The NTSB can also mandate certain actions. Before I decided to become a geologist, I was a commercial pilot flying small aircraft. Every few months I would read the latest NTSB reports. They made fascinating reading. As a young pilot, I wanted to make sure I didn't make the same mistakes as the unfortunate few highlighted by the NTSB. You might be surprised how many aircraft accidents (both large planes and small) were caused by easily avoidable things like running out of fuel, forgetting to put the landing gear down, or taking off with an overloaded aircraft.

Frank Nutter, President of the Reinsurance Association of America, and Dr. William Hooke, Director of the Policy Program at the American Meteorological Society, have a novel idea for assessing and improving the nation's resiliency to hazards. They think we should establish an organization like the NTSB to investigate damage to infrastructure after major natural disasters. I'd like to see such a National Flood Safety Board (NFSB), like the NTSB, have the power to mandate certain actions.

Let's use the flood example again, where local zoning decisions have such a big impact on future costs borne by the nation. The NFSB would be empowered to overrule decisions by local officials, who can be influenced by local politics, and who sometimes rubber-stamp applications for building in sensitive coastal zones. This would allow common sense actions that are often avoided, such as rebuilding in areas or ways less susceptible to future flood damage, or restoring coastal dunes and fringing vegetation to reduce future storm surge.

## THE CHALLENGE OF LARGE INFRASTRUCTURE PROJECTS

Large infrastructure projects are expensive. Billions of dollars were spent upgrading New Orleans' levees after Hurricane Katrina. Since it was well known that the levees would fail in a major hurricane, it made more sense to upgrade the levees *before* the hurricane. But the huge sums of money involved made that politically difficult – until the hurricane struck and politicians needed to be seen to be doing something useful.

Some classes of infrastructure are politically easier to fund than others. Every politician wants their name on new or upgraded schools and hospitals; fewer want their name on a new sewage treatment plant.

I lived in the city of Miami for many years. Shortly after I arrived in the early 1990s, while driving on a back street on a sunny day,

I came to a large body of water in the middle of the street. Where was the water coming from? And how deep was it? In my indecision, I stopped the car at the water's edge. A well-dressed woman behind me in a fancy car honked her horn and pulled around me, middle finger raised. She charged through the water, but got stuck half way across, the front of her vehicle nosing down into a depression. She opened her door and promptly fell into two feet of brown water, ruining her white dress. It turned out that a sewage line had ruptured under the street, making a sinkhole. Problems with sewage infrastructure were chronic then in Miami, as maintenance departments were starved for funds. Similar problems may become more common in the future as sea level rises and groundwater becomes more saline, degrading older steel and cast iron pipes.

Richard Elelman, a political analyst, former politician and current Head of Public Administration Projects in Catalonia, Spain, has thought deeply about these problems. He has an idea that would make it easier for politicians to lead in infrastructure planning and construction, especially on things like wastewater treatment and sewage lines, which can take many years to complete, counting design and construction. An independent board would outline the various steps needed to see the project through to completion. The board would score politicians on how well they funded their increment and moved the project to the next level. Credit would accrue for meeting each milestone, not just for project completion. Politicians who failed to move the ball forward on their watch would get a low score, and the public would be told this before the next election. Independent boards would have the expertise, time and mandate to resolve technically complex issues on long-term infrastructure projects that are so hard to address in the political arena.

## SUMMARY

This book has emphasized the need to "fit in" to our planet, with infrastructure and lifestyle that is both resilient in the face of hazards (both natural and human-caused), and sensitive to our ecosystem.

I have focused on problems, but have also tried to present commonsense solutions. A summary list appears after this chapter. Some of the entries on the list are laughably obvious, but people keep making the same mistakes, so I felt compelled to include the items. For instance: "Don't use fireworks inside a building; in the event of fire, open the doors so people can escape." In the last two decades, failure to follow this advice has caused hundreds of deaths in half a dozen countries.

Solving difficult environmental problems requires good communication between scientists and engineers on one hand, and the public and government officials on the other. Public media plays an essential role. Journalists and reporters need to publicize funding sources and conflicts of interest for pundits, and deny media exposure to those who refuse to disclose. As one of my colleagues said, "Bulls**t baffles brains." The media must become better at separating the former from the latter. More transparency at all levels of society will make it easier for reporters to do their job, and help people reach informed decisions.

How we deal with the interlinked problems of cost, pollution, sea-level rise and climate change associated with our energy and transportation infrastructure will be a huge challenge. Ending fossil fuel subsidies, implementing a carbon tax, researching new energy and transportation technologies, and changing our life style and infrastructure to low carbon alternatives will all help. Theses adjustments will be costly in the short-term, but will be more costly if we wait. I'll close with the hope that some of the ideas offered in this book can help to solve these problems, leaving our planet a better place for our grandchildren, and their grandchildren.

# Summary of Recommendations by Chapter

## CHAPTER 3

**3a. The importance of place.** Most earthquakes, tsunamis, volcanic eruptions, and flood events (both over-bank river flooding and hurricane-induced storm surge) happen in well-defined locations, for well-understood reasons. Zoning codes (where to build) and building codes (how to build) need to be developed and enforced. For construction that happened prior to scientific knowledge of particular risk, buildings can be upgraded as they approach the end of their economic life-span, but there should be transparency as to upgrade status, e.g. through websites or building plaques, to allow people to make informed choices.

**3b. Lack of scientific consensus.** In cases where the science is incomplete or scientists disagree, implement lower cost mitigation options, consistent with the lower threat estimates, until such time as scientific consensus is reached. There will always be some level of disagreement and uncertainty – don't use it as an excuse to do nothing.

**3c. Building fires.** Write and enforce strong building codes for fire safety, including multiple exits and the use of alarms and automatic sprinkler systems. Don't use fireworks inside a building. In the event of fire, open the doors so people can escape.

## CHAPTER 4

**4a. Beware of subduction zones.** Given current knowledge of earthquake and tsunami hazard, all subduction zones should be considered capable of generating magnitude 9 earthquakes and massive tsunamis. Assessment of earthquake and tsunami risk needs to consider events spanning at least the last few thousand years.

**4b. Think in 4-D (3-D plus time).** Related to 4a above. Nuclear power plant designers need to think about elevation as well as the horizontal position on their plant, and how geography and sea level could change in the future.

**4c. Back-up power.** Critical facilities in subduction zones, such as nuclear power plants, hospitals, etc., need to have backup power and control facilities on high ground where they are safe from flooding. Realistic stress tests need to be conducted regularly to insure reliable operation of backup hardware, software, procedures, and personnel.

**4d. Hydrogen gas.** Nuclear power plants need to have reliable, functioning passive hydrogen gas reactor/catalyst systems to prevent hydrogen explosions.

**4e. Radiation sensors.** Nuclear power plants need to have reliable, functioning radiation sensors to measure radiation leaks, including high dynamic range sensors for large leaks, and sensors at distant locations. Sensors should operate 24/7, and their data should be publicly available in real-time, accessible on the web and by smart phone. Background radiation measurements over broad regions should be taken prior to the emergency, not just in the atmosphere, but also in groundwater and, for plants in coastal areas, the coastal ocean.

**4f. Environmental monitoring during accidents.** In the event of a accident at a nuclear plant or a major chemical spill at an industrial facility, plans need to be in place to allow rapid access for environmental monitoring by qualified, objective third parties, including the academic community and NGOs.

**4g. Independent oversight.** Governments should strive to maintain independent oversight of sensitive or hazardous industries. Publicly available procedures and rules should be in place to address the "revolving door" problem between regulators and the regulated. These procedures should include a minimum separation time between employment on either side of the regulatory "fence." In addition, declarations of all potential conflicts of interest and past employment should be required and publicly available. Safety reviews should be

open to the public. Qualified media should be given access to design material and operating procedures to allow independent assessment.
**4h. Onshore vs offshore.** Sensitive coastal facilities such as nuclear power plants need to consider offshore as well as onshore hazards, and long-term changes in those hazards.

## CHAPTER 5

**5a. Earthquake and tsunami preparedness.** Pay attention to geological studies of rare long-interval events, and prepare accordingly.
**5b. Dangerous or explosive chemicals.** Storage facilities for dangerous chemicals and explosives should be designed and sited safely (not too close together for anything flammable or explosive; double-walled tanks for dangerous chemicals) and regularly inspected.

## CHAPTER 6

**6a. Follow the money.** The media have a critical role to play in reporting environmental hazards, and need to be keenly aware of the ability of vested interests to bias the public discourse. Similar advice is stated in Chapters 8 and 9, emphasizing its importance.
**6b. Mercury.** We need better ways of dealing with the insidious problem of mercury pollution. A tax or deposit on consumer items containing mercury such as batteries and compact fluorescent light bulbs should be considered. Heavy taxes should be levied on emissions. Seafood companies should measure and label their products as to mercury content, and they should work with environmental groups to reduce all sources of mercury pollution.
**6c. Fess up quickly.** Companies that have caused environmental problems should quickly "go public," working in an open, transparent way to fix problems and invest in long-term solutions. Companies that act promptly and in good faith should be rewarded with reduced penalties. Companies that don't should be subject to harsh penalties. If fatalities have resulted from specific actions of company officials, those officials should be subject to criminal as well as civil penalties.

**6d. Make reports readable.** Scientific or engineering committees who are studying an environmental problem should complete their report and then hand it to professional communicators who can make the report understandable to a non-technical audience.

**6e. Coal plants: Clean 'em or close 'em**. Coal-fired power plants should be phased out unless they are able to quickly adopt CCS or other technology to limit airborne pollutants. Older, highly polluting plants are generally not suitable for CCS. They have been allowed to operate for far too long and should be shut down. Workers at these plants should be eligible for retraining and public assistance to ease the transition to alternate employment.

**6f. Air pollution.** We need more focus on air pollution, our largest environmental killer. The various implicit and explicit policies that favor use of fossil fuels, a major source of air pollution, need to be changed. In particular, subsidies and tax incentives that promote fossil fuel use should be quickly phased out. They make no sense.

**6g. Promote renewables.** Invest in research to make solar panels more efficient and remove barriers that limit expansion of renewable energy (Florida, please take note).

## CHAPTER 7

**7a. High rainfall and over-bank river flooding.** Reduce "hardscape" (extended areas of asphalt and concrete) in cities, protect wetlands from development, and reduce development on river floodplains.

**7b. Coastal areas subject to hurricane-induced storm surge.** Reduce development in vulnerable low-lying coastal areas. Encourage expansion of coastal green space. Promote natural flood barriers that trap sediment and can survive rising sea level, such as mangroves, beach dunes, oyster beds, and sea grass.

**7c. Move to or make higher ground.** The cost of moving to higher ground should be subsidized for low-income residents. New and renovated construction in flood-prone areas should be "flood-proofed" by elevating the structure.

**7d. Levees.** Resurvey elevations periodically to verify height and health. Surveys need to be conducted with accurate satellite techniques to get around the problem of reference point subsidence.

**7e. Hospitals and other critical infrastructure** in flood-prone areas should have backup power and other sensitive equipment located above the first floor.

**7f. Plan for the new normal.** Mid-latitude cities that have not experienced hurricanes in the past need to plan for the occasional "extra-tropical cyclone" (translation – a hurricane where it's not supposed to be). Hurricanes are fueled by warm ocean water, and the oceans are warming, expanding the region of vulnerability.

**7g. Retreating from the coast.** Rising sea levels will eventually make some coastal areas untenable for continued occupation. For most areas this is many decades in the future, but it is inevitable. We should use the intervening time to plan a managed retreat. Changes to tax policy and insurance standards could help to minimize economic disruption. Subsidies for coastal development should be gradually reduced. Markets can play a useful role if left unimpeded, correctly assigning prices and factoring in issues such as insurance rates and the economic lifetime of buildings.

**7h. Critical coastal facilities** in danger of flooding that are not easily moved such as nuclear power plants need to plan for the occasional catastrophic flood.

## CHAPTER 8

**8a. Follow the money.** The media needs to be more aggressive at detailing the role of special interests in obfuscating the danger of fossil fuels and global warming.

**8b. Transparency.** Closely related to 8a above; who is funding what?

**8c. Political discourse and the media.** When politicians express viewpoints on scientific or environmental issues, experts who actually know something about the topic should be given equal time by the media. A license to broadcast should not be a license to broadcast drivel. The public has a right to know the facts. The media has a responsibility to deliver them.

CHAPTER 9

**9a. Improving scientific communication.** Scientists can improve their public communication with more use of images and videos. Reports for public consumption should have short and clearly written executive summaries, and be prepared with the help of professionals in communications and graphics.

**9b. Follow the money.** Media reporting of political speech needs to include follow-up on funding sources and potential conflicts of interest when politicians express views on scientific, environmental, and natural hazard issues.

**9c. Whistle blower lawsuits.** Scientists expressing professional opinions on environmental issues need to be protected from harassment by individuals and groups, including those funded by special interests. Laws need to be enacted to better protect these scientists, and enable prosecution of individuals, groups and business interests engaging in such harassment. Political ethics committees need to be more aggressive at prosecuting politicians who abuse the power of their office to harass scientists and scientific organizations that are simply doing their job.

**9d. Transparency: Building safety.** In seismic zones, the public should be informed of the ability of buildings to withstand earthquakes. Special plaques are one way to do this, and they should be prominently displayed.

**9e. Transparency: Environmental Audits.** Companies engaged in activities that pose environmental dangers should be subject to annual audits by an independent group to verify compliance with the latest safety and environmental standards. Results of the audit should be made public.

**9f. Disaster Bonds.** Companies engaged in activities that pose environmental dangers should be required to post bonds sufficient to cover the cost of the maximum plausible accident.

**9g. Special taxes** should be leveled on the production and use of elements and compounds that are damaging to public health, such as mercury and gold.

**9h. New tech and carbon taxes.** New technology is required in the areas of power generation and transportation, to reduce our consumption of fossil fuel, reduce $CO_2$ emissions, and reduce air pollution, which annually kills millions of people. The cost of the necessary research and development could be covered by a carbon tax.

**9i. Long-term economic development in Earth's tropical and subtropical belts.** Provide aid to poor countries for local research and development into clean energy sources.

**9j. Independent boards** should be established to oversee a variety of infrastructure needs, including those related to energy and the environment. Among their many duties, these boards could "score" politicians on the implementation of long-term infrastructure projects.

**9k. After-action reports.** After a major disaster, a committee of experts should be empowered to review all aspects of the disaster, including prior preparation, immediate disaster response, and subsequent recovery efforts. The resulting "lessons learned" report should be published, to serve as a blueprint for future mitigation efforts.

**9l. Establish NFSBs.** Countries that suffer frequent flood disasters should establish National Flood Safety Boards that are empowered to make zoning and building regulations in flood zones.

# References and Further Reading

*Entries marked with an asterisk have comments by the author.

## CHAPTER 1

Lott, N., and T. Ross (2006) Tracking and evaluating US billion dollar weather disasters, 1980–2005. NOAA/NCDC Technical Report: Asheville, NC.

Munich Re (2014) Topics GEO: Natural catastrophes 2013. Munich Re: München, Germany.

Nestle, M. (2015) *Soda Politics: Taking on Big Soda (and Winning)*. Oxford University Press: Oxford, UK.

Protti, M., V. González, A. V. Newman, T. H. Dixon, S. Y. Schwartz, J. S. Marshall, L. Feng, J. I. Walter, R. Malservisi, and S. E. Owen (2013) Prior geodetic locking resolved the rupture area of the anticipated 2012 Nicoya Earthquake. *Nature Geoscience*, 7, 117–121, doi:10.1038/ngeo2038.

Strunk, W., Jr., and E. B. White (1959) *The Elements of Style*. MacMillan: New York.

Taleb, N. N. (2010) *The Black Swan: The Impact of the Highly Improbable*. Random House: New York.

## CHAPTER 2

Bhattacharya, S. (2003) European heat wave caused 35,000 deaths. *New Scientist*, 10 October.

Chaline, E. (2013) *History's Worst Disasters*. New Burlington Books: London.

Chase, T. N., K. Wolter, R. A. Pielke Sr., and I. Rasool (2006) Was the 2003 European summer heat wave unusual in a global context? *Geophysical Research Letters*, 33, L23709, doi: 10.1029/2006GL027470.

Keller, R. C. (2015) *Fatal Isolation: The Devastating Paris Heat Wave of 2003*. University of Chicago Press: Chicago.

*Kolbert, E. (2014) *The Sixth Extinction*. Holt and Company: New York. An engaging description of the mass extinction of species, caused by habitat loss and environmental degradation that is currently underway.

Reid, H. F. (1910) *The Mechanics of the Earthquake, The California Earthquake of April 18, 1906, Report of the State Investigation Commission, Vol. 2*. Carnegie Institution of Washington: Washington, DC.

Robine, J.-M., S. L. K. Cheung, S. L. Roy, H. V. Oyen, C. Griffiths, J.-P. Michel, and F. R. Herrmann (2008) Death toll exceeded 70,000 in Europe during the summer of 2003. *Comptes Rendus Biologies*, 331, 2, 171–178.

Sella, G. F., T. H. Dixon, and A. Mao (2002) REVEL: A model for recent plate velocities from space geodesy. *Journal of Geophysical Research*, 107, B4, doi: 10.1029/2000JB000033.

Sigma (2014) *Natural Catastrophes and Man-Made Disasters in 2013*. Swiss Re: Zurich.

Stein, S. (2010) *Disaster Deferred: A New View of Earthquake Hazards in the New Madrid Seismic Zone*, Columbia University Press: New York.

Stern, R. J., D. W. Scholl, and G. Fryer (2016) *An Introduction to Convergent Plate Margins and their Natural Hazards*. American Geophysical Union/Wiley Blackwell: Washington, DC.

## CHAPTER 3

Boschi, E. (2013) L'Aquila's aftershocks shake scientists. *Science*, 341, 1451.

Calais, E., A. Freed, G. Mattioli, S. Jonsson, F. Amelung, P. Jansma, S.-H. Hong, T. H. Dixon, C. Prepetit, and R. Momplaisir (2010) Transpressional rupture of an unmapped fault during the 2010 Haiti earthquake. *Nature Geosciences*, doi: 10.1038/NGEO992.

Dixon, T. H., F. Farina, C. DeMets, P. Jansma, P. Mann, and E. Calais (1998) Relative motion between the Caribbean and North American plates based on a decade of GPS observations. *Journal of Geophysical Research*, 103, 15157–15182.

Hall, S. S. (2011) At fault? *Nature*, 477, 264–269.

Hough, S. E., J. Armbruster, L. Seeber, and J. Hough (2000) On the Modified Mercalli Intensities and magnitudes of 1811–1812 New Madrid earthquakes. *Journal of Geophysical Research*, 115, 23839–23864.

Jackson, D. D., K. Aki, C. A. Cornell, J. H. Dieterich, T. L. Henyey, M. Mahdyiar, D. Schwartz, and S. N. Ward (1995) Seismic hazards in southern California: Probable earthquakes, 1994 to 2024. *Bulletin of the Seismoloogical Society of America*, 85, 379–439.

Johnston, A. C. (1996) Seismic moment assessment of earthquakes in stable continental regions, III: New Madrid, 1811–1812, Charleston, 1886, and Lisbon, 1755. *Geophysical Journal International*, 126, 314–344.

Kolbe, A. R., R. A. Hutson, H. Shannon, E. Trzcinski, B. Miles, N. Levitz, M. Puccio, L. James, J. R. Noel, and R. Muggah (2010) Mortality, crime and access to basic needs before and after the Haiti earthquake: A random survey of Port-au-Prince households. *Medicine, Conflict, Survival*, 26, 281–297.

Manaker, D. M., E. Calais, A. M. Freed, S. T. Ali, P. Przybylski, G. Mattioli, P. Jansma, C. Prepetit, and J. B. De Chabalie (2008) Interseismic plate coupling and strain partitioning in the Northeastern Caribbean. *Geophysical Journal International*, 174, 889–903.

Pritchard, C. (2012) L'Aquila ruling: Should scientists stop giving advice? *BBC News Magazine*, October 28.

Reid, H. F. (1910) *The Mechanics of the Earthquake, The California Earthquake of April 18, 1906, Report of the State Investigation Commission, Vol. 2*. Carnegie Institution of Washington: Washington, D.C.

Ropeik, D. (2012) The L'Aquila verdict: A judgement not against science, but against a failure of science communication. *Scientific American*, Guest Blog, October 22.

*Scherer, J. (1912) Great earthquakes in the island of Haiti. *Bulletin of the Seismological Society of America*, 2, 161–180. One of the earliest warnings concerning earthquake hazards for Haiti.

Schumpeter, J. (1942) *Capitalism, Socialism, and Democracy*. Harper: New York.

Stein, S. (2010) *Disaster Deferred: A New View of Earthquake Hazards in the New Madrid Seismic Zone*. Columbia University Press: New York.

Stein, S., and J. Stein (2013) *Playing against Nature: Integrating Science and Economics for Cost-Effective Natural Hazard Mitigation*. John Wiley & Sons: Chichester, UK.

Sykes, L. R., W. R. McCann, and A. L. Kafka (1982) Motion of the Caribbean Plate during the last 7 million years and implications for earlier Cenozoic movements. *Journal of Geophysical Research*, 87, 10656–10676.

Taber, S. (1922) The great fault troughs of the Antilles. *Journal of Geology*, 30, 89–114.

Weldon, R., K. Scharer, T. Fumal, and G. Biasi (2004) Wrightwood and the earthquake cycle: What the long recurrence record tells us about how faults work. *GSA Today*, 14, 4–10, doi: 10.1130/1052–5173(2004)014.

Weldon, R. J., T. E. Fumal, G. P. Biasi, and K. M. Scharer (2005) Past and future earthquakes on the San Andreas Fault. *Science*, 308, 966–967.

## CHAPTER 4

*Bailly du Bois P., P. Laguionie, D. Boust, I. Korsakissok, D. Didier, and B. Fiévet (2012) Estimation of marine source-term following Fukushima Dai-ichi accident. *Journal of Environmental Radioactivity*, 114, 2–9. An important reference describing measurement of the marine contamination from Fukushima.

Buesseler, K. O. (2012) Fishing for answers off Fukushima. *Science*, 338, 480–482.

Buesseler, K. O., S. R. Jayne, N. S. Fisher, I. I. Rypina, H. Baumann, Z. Baumann, C. F. Breier, E. M. Douglass, J. George, A. M. Macdonald, H. Miyamoto, J. Nishikawa, S. M. Pike, and S. Yoshida (2012) Fukushima-derived radionuclides in the ocean and biota off Japan. *Proceedings of the National Academy of Sciences*, 109, 5984–5989.

Charette, M. A., C. F. Breier, P. B. Henderson, S. M. Pike, I. I. Rypina, S. R. Jayne, and K. O. Buesseler (2013) Radium-based estimates of cesium isotope transport and total direct ocean discharges from the Fukushima nuclear power plant accident. *Biogeosciences*, 10, 2159–2167, doi:10.5194/bg-10-2159-2013.

Connor, C. B. (2011) A quantitative literacy view of natural disasters and nuclear facilities. *Numeracy*, 4, 2, Article 2, doi: http://dx.doi.org/10.5038/1936-4660.4.2.2.

Daniell, J.E., A. Vervaeck, and F. Wenzel, (2011) A timeline of the socio-economic effects of the 2011 Tohoku earthquake with emphasis on the development of a new worldwide rapid earthquake loss estimation procedure. *Australian Earthquake Engineering Society Conference*, November.

Dvorak, P., and P. Landers (2011) Japanese plant had barebones risk plan. *Wall Street Journal*, March 31.

Fackler, M. (2011a) Japan: 3 damaged reactors may take 40 years to decommission, officials say. *New York Times*, December 22, A15.

Fackler, M. (2011b) More radioactive water leaks at Japanese plant. *New York Times*, December 5, A11.

Fackler, M. (2011c) Tsunami warnings for the ages, carved in stone. *New York Times*, April 21, A6.

Fackler, M. (2012). Japan power company admits failings on plant precautions. *New York Times*, October 13, A4.

Fackler, M. (2015) Japan: Leak is disclosed at nuclear plant. *New York Times*, February 25, A8.

Funabashi, Y., and Kay Kitazawa (2012) Fukushima in review: A Complex disaster, a disastrous response. *Bulletin of the Atomic Scientists*, 68, March/April, 13–14.

Geller, R. J. (2011) Shake-up time for Japanese seismology. *Nature*, 472, 407–409.

Gilhooly, R. (2011) Quake prediction myth debunked: Professor's new book blows cover off panic-driven ¥10 trillion industry. Special to *Japan Times*, December 27.

Kanamori, H. (1972) Mechanism of tsunami earthquake. *Physics of the Earth and Planetary Interiors*, 6, 346–359.

Keefe, P. R. (2007) The Jefferson bottles. *The New Yorker*, September 3.

Kiger, P. J. (2013) Fukushima's radioactive water leak: What you should know. *National Geographic News*, August 9.

Kim, V. (2011) Japan damage could reach $235 billion, World Bank estimates. *Los Angeles Times*, March 21.

McCaffrey, R. (2007) The next great earthquake. *Science*, 315, 1675–1676.

McCaffrey, R. (2008) Global frequency of magnitude 9 earthquakes. *Geology*, 36, 263–266.

McCalpin, J. (2009) *Paleoseismology*, 2nd edn. Academic Press: Burlington, MA.

Minoura K., and Nakaya S. (1991) Traces of tsunami preserved in inter-tidal lacustrine and marsh deposits: Some examples from northeast Japan. *Journal of Geology*, 99, 265–287.

Minoura, K., F. Imamura, D. Sugawara, Y. Kono, and T. Iwashita (2001) The 869 Jogan tsunami deposit and recurrence interval of large-scale tsunami on the Pacific coast of northeast Japan. *Journal of Natural Disaster Science*, 23, 83–88.

Nöggerath, J., R. J. Geller, and V. K. Gusiakov (2011) Fukushima: The myth of safety, the reality of geoscience. *Bulletin of the Atomic Scientists*, 67, 37–46.

Rypina, I. I., S. R. Jayne, S. Yoshida, A. M. Macdonald, E. Douglass, and K. Buesseler (2013) Short-term dispersal of Fukushima-derived radionuclides off Japan: Modeling efforts and model-data intercomparison. *Biogeosciences Discussion*, 10, 1517–1550, doi: 10.5194/bgd-10-1517-2013.

Sieh, K. E. (1978) Prehistoric large earthquakes produced by slip on the San Andreas Fault at Pallett Creek, California. *Journal of Geophysical Research*, 83, 3907–3939.

Stein, J. L., and S. Stein (2014) Gray swans: Comparison of natural and financial hazard assessment and mitigation. *Natural Hazards*, 72, 1279–1297.

Stein, S., and E. Okal (2005) Seismology: Speed and size of the Sumatra earthquake. *Nature*, 434, 581–582, doi: 10.1038/434581a

Stein, S., and E. Okal (2007) Ultralong period seismic study of the December 2004 Indian Ocean earthquake and implications for regional tectonics and the subduction process, *Bull. Seism. Soc. Am., 97, S279–S295.*

Stohl, A., P. Seibert, G. Wotawa, D. Arnold, J. F. Burkhart, S. Eckhardt, C. Tapia, A. Vargas, and T. J. Yasunari (2012) Xenon-133 and caesium-137 releases into the atmosphere from the Fukushima Dai-ichi nuclear power plant: Determination of the source term, atmospheric dispersion, and deposition. *Atmospheric Chemistry and Physics*, 12, 2313–2343.

UNSCEAR (2008) *Sources and Effects of Ionizing Radiation, UN Scientific Committee on the Effects of Atmospheric Radiation.* United Nations: New York.

Wallace, B. (2008) *The Billionaire's Vinegar: The Mystery of the World's Most Expensive Bottle of Wine*. Random House: New York.

Watanabe H. (1998) *Comprehensive List of Tsunamis that Struck the Japanese Islands*, 2nd edn. (in Japanese). Tokyo: University of Tokyo Press.

Yamaguchi, M. (2012) Japan utility agrees nuclear crisis was avoidable. *Associated Press*, October 12.

Yeats, R. (2012) *Active Faults of the World*. Cambridge University Press: Cambridge, UK.

## CHAPTER 5

Adams, J. (1984) Active deformation of the Pacific northwest continental margin. *Tectonics*, 3, 449–472.

Adams, J. (1990) Paleoseismicity of the Cascadia subduction zone: Evidence from turbidites off the Oregon-Washington margin. *Tectonics*, 9, 569–583.

Ansal, A. A. Akinci, G. Cultrera, M. Erdik, V. Pessina, G. Tönük, and G. Ameri (2009) Loss estimation in Istanbul based on deterministic earthquake scenarios of the Marmara Sea region (Turkey). *Soil Dynamics and Earthquake Engineering*, 29, 699–709.

Atwater B. F. (1987) Evidence for great Holocene earthquakes along the outer coast of Washington State. *Science*, 236, 942–944.

Atwater, B. F., and D. K. Yamaguchi (1991) Sudden, probably co-seismic submergence of Holocene trees and grass in coastal Washington State. *Geology*, 19, 706–709.

Atwater, B. F., S. Musumi-Rokkaku, K. Satake, Y. Tsuji, K. Ueda, and D. K. Yamaguchi (2005) *The Orphan Tsunami of 1700*. University of Washington Press: Seattle.

Barka, A. (1996) Slip distribution along the North Anatolian fault associated with the large earthquakes of the period 1939–1967. *Seismological Society of America Bulletin*, 86, 1238–1254.

Bouma, A. H. (1962) *Sedimentology of Some Flysch Deposits*. Elsevier: Amsterdam.

Erdik, M. (2013) Earthquake risk in Turkey. *Science*, 341, 724–725.

Goldfinger, C., C. H. Nelson, and J. E. Johnson (2003) Holocene earthquake records from the Cascadia subduction zone and northern San Andreas fault based on precise dating of offshore turbidites. *Annual Reviews of Earth and Planetary Science*, 31, 555–577.

Goldfinger, C., C. H. Nelson, A. Morey, J. E. Johnson, J. Gutierrez-Pastor, A. T. Eriksson, E. Karabanov, J. Patton, E. Gracia, R. Enkin, A. Dallimore, G. Dunhill, and T. Vallier (2012) *Turbidite Event History: Methods and Implications for Holocene Paleoseismicity of the Cascadia Subduction Zone. USGS Professional Paper 1661-F*. U.S. Geological Survey: Reston, VA, http://pubs.usgs.gov/pp/pp1661f/.

Goldfinger, C., Y. Ikeda, R. S. Yeats, and J. Ren (2013) Superquakes and supercycles. *Seismological Research Letters*, 84, 24–32.

Griggs, G. B., and L. D. Kulm (1970) Sedimentation in Cascadia deep-sea channel. *Geological Society of America Bulletin*, 81, 1361–1384, doi: 10.1130/0016-7606(1970)81[1361: SICDC]2.0.CO;2.

Heaton. T. H., and H. Kanamori (1984) Seismic potential associated with subduction in the northwestern United States. *Seismological Society of America Bulletin*, 74, 933–941.

Heezen, B. C., and M. Ewing (1952) Turbidity currents and submarine slumps, and the 1929 Grand Banks earthquake. *American Journal of Science*, 250, 849–73

Houtz, R. E., and H. W. Wellman (1962) Turbidity Current at Kadavu Passage, Fiji. *Geological Magazine*, 99, 57–62.

Jacoby, G. C., D. E. Bunker, and B. E. Benson (1997) Tree-ring evidence for an A.D. 1700 Cascadia earthquake in Washington and northern Oregon. *Geology*, 25, 999–1002.

Kastens, K. A. (1984) Earthquakes as a triggering mechanism for debris flows and turbidites on the Calabrian Ridge. *Marine Geology*, 55, 13–33.

Ketin, I. (1948) Uber die tektonisch.mechanischen Folgerungen aus den groflenanatolischen Erdbeben des letzten Dezenniums. *Geologische Rundschau*, 36, 77–83.

Kremer, K., G. Simpson, and S. Girardclos (2012) Giant Lake Geneva tsunami in AD 563. *Nature Geoscience* 5, 756–757, doi:10.1038/ngeo1618.

Kremer, K., F. Marillier, M. Hilbe, G. Simpson, D. Dupuy, B. J. F. Yrro, A.-M. Rachoud-Schneider, P. Corboud, B. Bellwald, W. Wildi, and S. Girardclos (2014) Lake dwellers occupation gap in Lake Geneva (France–Switzerland) possibly explained by an earthquake–mass movement–tsunami event during Early Bronze Age. *Earth and Planetary Science Letters*, 385, 28–39.

McClusky, S., S. Balassanian, A. Barka, C. Demir, S. Ergintav, I. Georgiev, O. Gurkan, M. Hamburger, K. Hurst, H. Kahle, K. Kastens, G. Kekelidze, R. King, V. Kotzev, O. Lenk, S. Mahmoud, A. Mishin, M. Nadariya, A. Ouzounis, D. Paradissis, Y. Peter, M. Prilepin, R. Reilinger, I. Sanli, H. Seeger, A. Tealeb, M. N. Toksöz, and G. Veis (2000) Global Positioning System constraints on plate kinematics and dynamics in the eastern Mediterranean and Caucasus. *Journal of Geophysical Research*, 105, 5695–5719.

Morner, N. A. (1985) Paleoseismicity and geodynamics in Sweden. *Tectonophysics*, 117, 139–153.

Parsons, T. (2004) Recalculated probability of M>7 earthquakes beneath the Sea of Marmara, Turkey. *Journal of Geophysical Research*, 109, B05304, doi: 10.1029/2003JB002667.

Pilkey, O. H. (1988) Basin Plains: Giant sedimentation events. *Geological Society of America*, Special Paper 229, 93–99.

Pondard, N., R. Armijo, G. C. P. King, B. Meyer, and F. Flerit (2007) Fault interactions in the Sea of Marmara pull-apart (North Anatolian Fault): Earthquake clustering and propagating earthquake sequences. *Geophysical Journal International*, 171, 1185–1197.

Richter, C. (1958) *Elementary Seismology*, 1st edn. W. H. Freeman and Co.: San Francisco.

Satake, K, K. Shimazaki, Y. Tsuji, and K. Ueda (1996) Time and size of a giant earthquake in Cascadia inferred from Japanese tsunami records of January, 1700. *Nature*, 379, 246–249.

Savage, J. C., M. Lisowski, and W. H. Prescott (1981) Geodetic strain measurements in Washington. *Journal of Geophysical Research*, 86, 4929–4940.

Sims, J. D. (1975) Determining earthquake recurrence intervals from deformational structures in young lacustrine sediments. *Tectonophysics*, 29, 141–152.

Stein, R. S., A. A. Barka, and J. H. Dieterich (1997) Progressive failure on the North Anatolian fault since 1939 by earthquake stress triggering. *Geophysical Journal International*, 128, 594–604.

Thompson, J. (2012) *Cascadia's Fault: The Coming Earthquake and Tsunami that Could Devastate North America*. Counterpoint: Berkeley, CA.

Uzel, T., K. Eren, and A. A. Dindar (2010) *Monitoring Plate Tectonics and Subsidence in Turkey by CORS-TR and InSAR. FIG Congress*: Sydney, Australia.

Yamaguchi, D.K., B. F. Atwater, D. E. Bunker, B. E. Benson, and M. S. Reid (1997) Tree-ring dating the 1700 Cascadia earthquake. *Nature*, 389, 922–923

Yeats, R. (2012) *Active Faults of the World*. Cambridge University Press: Cambridge, UK.

Yeats, R. (2015) *Earthquake Time Bombs*. Cambridge University Press: Cambridge, UK.

## CHAPTER 6

Brown, L. R., J. Larsen, J. M. Roney, and E. E Adams (2015) *The Great Transition: Shifting from Fossil Fuels to Solar and Wind Energy*. W. W. Norton and Co.: New York and London.

*Bryson, B. (2004), *A Short History of Nearly Everything*. Broadway Books: New York. His chapter "Getting the Lead Out" is an excellent summary of the history of leaded gasoline.

*Davidson, C. I., ed. (1999) *Clean Hands: Clair Patterson's Crusade against Environmental Lead Contamination*. Nova Science: Hauppauge, NY. A great read about a great person.

Deffeyes, K. (2001) *Hubbert's Peak: The Impending World Oil Shortage*. Princeton University Press: Princeton, NJ.

Dekok, D. (2009) *Fire Underground: The Ongoing Tragedy of the Centralia Mine Fire*, Globe Pequot Press: Guilford, CT.

*Denworth, L. (2009) *Toxic Truth: A Scientist, A Doctor, and the Battle over Lead*. Beacon Press: Boston. A comprehensive account of the battles over lead pollution.

Flight, D. M. A., and A. J. Scheib (2011) Soil geochemical baselines in the UK urban centres: the G-BASE project. In: *Mapping Chemical Environments in Urban Areas*, eds. C. C. Johnson, A. Demetriades, J. Locatura, and R. T. Otteson, Ch. 13. John Wiley and Sons: Chichester, UK.

Greenpeace (2006) *The Chernobyl Catastrophe: Consequences on Human Health*. Greenpeace: Netherlands.

Hodes, G. (2013) Perovskite-Based Solar Cells. *Science*, 342, 6156, 317–318, doi: 10.1126/science.1245473

Hightower, J. (2011) *Diagnosis Mercury: Money, Politics and Poison*. Island Press: Washington, DC.

Hofmeiseter, J. (2011) *Why We Hate the Oil Companies*. Palgrave/MacMillan: New York.

Kunstler, J. H. (2006) *The Long Emergency: Surviving the End of Oil, Climate Change, and Other Converging Catastrophes of the Twenty-First Century*. Grove Press: New York.

Lomburg, B. (2014) A report card for humanity: 1900–2050. *The Atlantic*, January.

Mason, R. P., A. L. Choi, W. F. Fitzgerald, C. R. Hammerschmidt, C. I. H. Lamborg, A. L. Soerensen, and E. M. Sunderland (2012) Mercury biogeochemical cycling in the ocean and policy implications. *Environmental Research*, 119, 101–117.

Mayer, J. (2015) *Dark Money: The Hidden History of the Billionaires behind the Rise of the Radical Right*. Doubleday: New York, Toronto.

McGrayne, S. B. (2001) *Prometheans in the Lab*. McGraw-Hill: New York.

Mielke, H. (1999) Lead in the inner-cities. *American Scientist*, 87, 62–73.

Murray, J. W. (2013) Peak oil and energy independence: myth and reality. *EOS: Transactions, American Geophysical Union*, 94, 245–246.

National Academy (2010) *Hidden Costs of Energy: Unpriced Consequences of Energy Production and Use*. National Academy Press: Washington, DC.

Needleman H. L. (2008) The case of Deborah Rice: Who is the Environmental Protection Agency protecting? *PLoS Biology*, 6, 5, 0940–0942, doi: 10.1371/journal.pbio.006012.

Needleman H. L., C. Gunnoe, A. Leviton, R. Reed, H. Peresie, C. Maher, and P. Barrett (1979) Deficits in psychological and classroom performance of children with elevated dentine lead levels. *New England Journal of Medicine*, 300, 689–695.

Needleman H. L., A. Schell, D. Bellinger, A. Leviton, and E. N. Allred (1990) The long-term effects of exposure to low doses of lead in childhood. An 11-year follow-up report. *New England Journal of Medicine*, 322, 83–88.

Needleman H. L., O. Tuncay, and I. M. Shapiro (1972) Lead levels in deciduous teeth of urban and suburban American children. *Nature*, 235, 111–112.

Nesheim, M. C., and A. L. Yaktine (2007) *Seafood Choices: Balancing Benefits and Risks*. National Academies Press: Washington, DC.

Oreskes, N., and E. M. Conway (2010) *Merchants of Doubt: How a Handful of Scientists Obscured the Truth on Issues from Tobacco Smoke to Global Warming*. Bloomsbury Press: New York, London.

Palus, S. (2015) Humans greatly increase mercury in the ocean. *EOS: Earth and Space Science News*, 96 (10), 28.

Patterson, C. (1956) Age of meteorites and the Earth. *Geochimica et Cosmochimica Acta* 10 (4), 230–237.

*Patterson, C. (1965) Contaminated and natural lead environments of man. *Archives of Environmental Health*, 11, 344–360. This paper was the first to show that high lead levels in industrial nations are both widespread and man-made.

Patterson, C., J. Ericson, M. Manea-Krichten, and H. Shirahata (1991) Natural skeletal levels of lead in *Homo sapiens sapiens* uncontaminated by technological lead. *Science of the Total Environment*, 107, 205–236.

Quiqley, J. (2007) *The Day the Earth Caved In: An American Mining Tragedy*. Random House: New York.

Ropeik, D. (2010) *How Risky is it Really? Why Our Fears Don't Always Match the Facts*. McGraw-Hill: New York.

Scheib, C., D. Flight, T. R. Lister, A. Scheib, and N. Breward (2011) Lead (Pb) in topsoil. G-BASE Geochemical Map, British Geological Survey.

Simon, S. L. (2002) Radiation doses to local populations near nuclear weapons test sites worldwide. *Health Physics*, 82, 706–725.

Simon, S. L., A. Bouville, and C. E. Land (2006) Fallout from nuclear weapons tests and cancer risks. *American Scientist*, 94, 48–57.

Streets, D. G., M. K. Devane, Z. Lu, T. C. Bond, E. M. Sunderland, and D. J. Jacob (2011) All-time releases of mercury to the atmosphere from human activities. *Environmental Science and Technology*, 45, 10485–10491.

UNSCEAR (2008) *Sources and effects of ionizing radiation, United Nations Scientific Committee on the Effects of Atomic Radiation, Report to the General Assembly with Scientific Annexes, Volume 2, Scientific Annexes C, D, and E.* United Nations: New York.

*Wall Street Journal* (1965) Controversy flares over amount of lead absorbed by humans. September 9, 11.

Wenar, L. (2015) *Blood Oil: Tyrants, Violence, and the Rules That Run the World.* Oxford University Press: Oxford, UK.

Xu, L. Q., X.-D. Liu, L. Sun, Q.-Q. Chen, H. Yan, Y. Liu, Y.-H. Luo, and J. Huang (2011) A 700-year record of mercury in avian eggshells of Guangjin Island, South China Sea. *Environmental Pollution*, 159, 889–896.

Zhang, Y. L. Jaegle, L. Thompson, and D. G. Streets (2014) Six centuries of changing oceanic mercury. *Global Biogeochemical Cycles*, 28, 1251–1261.

## CHAPTER 7

American Geophysical Union (2006) Hurricanes and the U.S. Gulf Coast: Science and Sustainable Rebuilding. AGU Special Report, Washington, DC.

American Society of Civil Engineers (ASCE) (2007) The New Orleans Hurricane Protection System: What Went Wrong and Why. A Report by the American Society of Civil Engineers Hurricane Katrina External Review Panel, Reston, VA.

Bonin, J., and D. Chambers (2013) Uncertainty estimates of a GRACE inversion modelling technique over Greenland using a simulation. *Geophysical Journal International*, 194, 212–229.

Chaussard, E., F. Amelung, H. Abidin, and S.-H. Hong (2013) Sinking cities in Indonesia: ALOS PALSAR detects rapid subsidence due to groundwater and gas extraction. *Remote Sensing of Environment*, 128, 150–161, doi:10.1016/j.rse.2012.10.015.

Dixon, T. H. (2015) Ten years after Katrina: What have we learned? *Eos, Earth and Space Science News*, 96, doi:10.1029/2015EO034703.

Dixon, T. H., and M.E. Parke (1983) Bathymetry estimates in the southern oceans from SEASAT radar altimetry. *Nature*, 304, 406–411.

Dixon, T. H., M. Naraghi, M. K. McNutt, and S. M. Smith (1983) Bathymetic prediction from SEASAT altimeter data. *Journal of Geophysical Research*, 88, 1563–1571.

Dixon, T. H., and R. K. Dokka (2008) Earth scientists and public policy: Have we failed New Orleans? *EOS: Transactions of the American Geophysical Union*, 89, 10, 96.

Dixon, T. H., F. Amelung, A. Ferretti, F. Novali, F. Rocca, R. Dokka, G. Sella, S.-W. Kim, S. Wdowinski, and D. Whitman (2006) New Orleans subsidence: Rates and spatial variation measured by permanent scatterer interferometry. *Nature*, 441, 587–588.

Fink, S. (2013) *Five Days at Memorial: Life and Death in a Storm-Ravaged Hospital.* Crown: New York.

Fu, L. L., and A. Cazenave (2001) *Satellite Altimetry and Earth Sciences. International Geophysics Series, v. 69.* Academic Press: New York, London, Tokyo.

Glass, A. and O. Pilkey (2013) *Denying Sea Level Rise: How 100 Centimeters Divided the State of North Carolina. Earth*, 58, 5, 26–33.

Interagency Performance Evaluation Task Force (IPET) (2007–2009), *Performance Evaluation of the New Orleans and Southeast Louisiana Hurricane Protection System: Final Report of the Interagency Performance Evaluation Task Force*, 9 vols. U.S. Army Corps of Engineers: Vicksburg, MS.

Karegar, M. A., T. H. Dixon, and R. Malservisi (2015) A three-dimensional surface velocity field for the Mississippi Delta: Implications for coastal restoration and flood potential. *Geology*, 43, 519–522.

Karegar, M. A., T. H. Dixon, and S. E. Engelhart (2016) Subsidence along the Atlantic coast of North America: Insights from GPS and late Holocene relative sea level data. *Geophysical Research Letters*, 43, doi: 10.1002/2016GL068015.

Larson, E. (2000) *Isaac's Storm: A Man, a Time, and the Deadliest Hurricane in History*. Vintage Books: New York, Toronto.

Ludden, J. (2013) A Hot Topic: Climate Change Coming to Classrooms. *Morning Edition*, March 27.

Machiavelli, N. (1532) *The Prince*, translated by H. C. Mansfield, Jr., 1985 edition, University of Chicago Press: Chicago.

Moore, M. (1989) Roger & Me. Documentary film on Roger Smith, former chairman of GM, by filmmaker Michael Moore. Preserved in the US National Film Registry, Library of Congress, in 2013.

Nocquet, J.-M., E. Calais, and B. Parsons (2005) Geodetic constraints on glacial isostatic adjustment in Europe. *Geophysical Research Letters*, 32, L06308.

Parris, A., P. Bromirski, V. Burkett, D. Cayan, M. Culver, J. Hall, R. Horton, K. Knuuti, R. Moss, J. Obeysekera, A. Sallenger, and J. Weiss (2012) *Global Sea Level Rise Scenarios for the US National Climate Assessment*. NOAA Tech Memo OAR CPO-1: Washington, DC.

Pfeffer, W. T., J. T. Harper, and S. O'Neel (2008) Kinematic constraints on glacier contributions to 21st-century sea level rise. *Science*, 321, 1340–1343.

Pilkey, O.H., and R. Young (2009) *The Rising Sea*. Island Press: Washington, DC.

Riggs, S. R., D. V. Ames, S. J. Culver, and D. J. Mallinson (2011) *The Battle for North Carolina's Coast*. University of North Carolina Press: Chapel Hill, NC.

Rohling, E. J., K. Grant, CH. Hembleen, M. Siddall, B. A. A. Hoogakker, M. Bolshaw, and M. Kucera (2008) High rates of sea level rise during the last interglacial period. *Nature Geoscience*, 1, 38–42.

Sella, G., S. Stein, T. H. Dixon, M. Craymer, T. James, S. Mazzotti, and R. K. Dokka (2007) Observation of glacial isostatic adjustment in "stable" North America with GPS. *Geophysical Research Letters*, 34, L02306, doi:10:1-29/2006GL027081.

Sigma (2009) *Natural Catastrophes and Man-Made Disasters in 2008*. Swiss Re: Zurich.

Tapley, B. D., S. Bettadpur, M. Watkins, and C. H. Reigber (2004) The gravity recovery and climate experiment: Mission overview and early results. *Geophysical Research Letters*, 31, L09607, doi:10.1029/2004GL019920.

Trenberth, K. (2012) Opinion: Super Storm Sandy: What role did climate change play in this week's massive hurricane. *The Scientist*, October 31.

Trenberth, K. E., C. A. Davis, and J. Fasullo (2007) Water and energy budgets of hurricanes: Case studies of Ivan and Katrina. *Journal of Geophysical Research*, 112, D23106, doi:10.1029/2006JD008303.

*US Army Corps of Engineers (2006) Performance Evaluation, Status and Interim Results, Report 2 of a Series, Performance Evaluation of the New Orleans and Southeast Louisiana Hurricane Protection System, Interagency Performance Evaluation Task Force (IPET). Note: a modified version of this report was published in 2009 by the National Academies Press, Washington, DC.

*Van Houten, C. (2016) The first official climate refugees in the U.S. race against time. National Geographic (published online, May 25). A poignant story of a Mississippi Delta community losing its land due to subsidence and sea level rise.

Wdowinski, S., R. Bray, B. P. Kirtman, and Z. Wu (2016) Increasing flooding hazard in coastal communities due to rising sea level: Case study of Miami Beach, Florida. *Ocean & Coastal Management*, 126, 1–8.

*Wendel, J. (2016) Dirty water: Unintended consequences of climate resiliency. *EOS: Earth and Space Science News*, 97, doi:10.1029/2016EO047061. An interesting story of what happens when you pump urban floodwater into a nearby bay (warning – its not pretty).

Wouters, B., D. Chambers, and E. J. O. Schrama (2008) GRACE observes small-scale mass loss in Greenland. *Geophysical Research Letters*, 35, L20501, doi:10.1029/2008GL034816.

Yang, Q., T. H. Dixon, P. G. Myers, J. Bonin, D. Chambers, and M. R. van den Broeke (2016) Recent increases in Arctic freshwater flux affects Labrador Sea convection and Atlantic overturning circulation. *Nature Communications*, 7, 10525, doi: 10.1038/ncomms10525.

## CHAPTER 8

*Arrhenius, S. (1896) On the influence of carbonic acid in the air upon the temperature of the ground. *Philosophical Magazine*, 41, 237–276. Carbonic acid is the old name for carbon dioxide.

*Associated Press* (2016) Migrant toll over 1,000 in a week: Deaths in the Mediterranean come amid a rise in bids to reach Europe. June 1.

Barnett, C. (2015) *Rain: A Natural and Cultural History*. Crown: New York.

Bowden, M. (1999) *Black Hawk Down: a Story of Modern War*. Grove Press: New York.

*Brandt, N. (2016) *Inherit the Dust*. Edwynn Houk Gallery: New York. A stark portrayal of environmental degradation and habitat loss, in pictures.

Chatterjee, R. (2016) Occupational hazard: Farm workers are dying in southern India as an epidemic of a mysterious kidney disease goes global. *Science*, 352, 24–27.

Cornwall, W. (2015) Ghosts of oceans past. *Science*, 350, 752–755

Davenport, C. (2013) The cost of inaction. *National Journal*, 45, 6, 8–13

*Diamond, J. (2004) *Collapse: How Societies Choose to Fail or Succeed*. Penguin: New York, Toronto. Fascinating studies of how and why various societies have failed through the ages (environmental problems are a surprisingly common cause).

*Dutton, A., A. E. Carlson, A. J. Long, G. A. Milne, P. U. Clark, R. DeConto, B. P. Horton, S. Rahmstorf, and M. E. Raymo (2015) Sea-level rise due to polar ice-sheet mass loss during past warm periods. *Science*, 349, doi: 10.1126/science.aaa4019-1 – 4019-9. This article shows that sea level was at least 6 meters higher in previous periods of earth history when atmospheric carbon dioxide concentrations were similar to today.

Ehrlich, P., and A. Ehrlich (2004) *One with Nineveh: Politics, Consumption and the Human Future*. Island Press/Shearwater: Washington, DC.

Foster, G. L., and E. J. Rohling (2013) Relationship between sea level and climate forcing by $CO_2$ on geological time scales. *Proceedings, National Academy of Sciences*, 110, 1209–1214.

Hammer, J. (2013) Is a lack of water to blame for the conflict in Syria? *Smithsonian Magazine*, 44, 3, 18.

Hansen, J., M. Sato, and R. Ruedy (2012) Perception of climate change. *Proceedings, National Academy of Science*, e2415–e2423, doi: 10.1073/pnas.1205276109.

Hassol, S. J. (2008) Improving how scientists communicate about climate change. *Eos: Transactions, American Geophysical Union*, 89, 11, 106–107.

Hoerling, M., A. Kumar, R. Dole, J. W. Nielsen-Gammon, J. Eischeid, J. Perlwitz, X-W. Quan, T. Zhang, P. Pegion, and M. Chen (2013) Anatomy of an extreme event. *Journal of Climate*, 26, 2811–2832.

Houghton, J. (2015) *Global Warming: The Complete Briefing*, 5th edn. Cambridge University Press: Cambridge, UK.

Hudson, V. M., B. Ballif-Spanvill, M. Caprioli, and C. F. Emmett (2014) *Sex and World Peace*. Columbia University Press: New York.

Keeling, C. D. (1960) The concentration and isotopic abundances of carbon dioxide in the atmosphere. *Tellus*, 12, 2, 200–203.

Kolbert, E. (2006) *Field Notes from a Catastrophe: Man, Nature, and Climate Change*. Bloomsbury Publishing: New York and London.

Kolbert, E. (2014) *The Sixth Extinction*. Henry Holt and Company: New York.

Kunstler, J. H. (2006) *The Long Emergency: Surviving the End of Oil, Climate Change, and Other Converging Catastrophes of the Twenty-First Century*. Grove Press: New York.

Lachapelle, E., S. Dinan, C. Borick, B. Rabe, and S. Mills (2015) *Mind the Gap: Climate Change Opinions in Canada and the United States*. Published jointly by the *National Survey on Energy and the Environment* (US) and *National Survey of Canadian Public Opinion* (Canada). Woodrow Wilson International Center for Scholars: Washington, DC.

Mann, M. E., R. S. Bradley, and M. K. Hughes (1998) Global-scale temperature patterns and climate forcing over the past six centuries. *Nature*, 392, 779–787.

Mann, M. E., R. S. Bradley, and M. K. Hughes (1999) Northern hemisphere temperatures during the past millennium: Inferences, uncertainties, and limitations. *Geophysical Research Letters*, 26, 759–762.

Mann, M. E., Z. Zhang, M. K. Hughes, R. S. Bradley, S. K. Miller, S. Rutherford, and F. Ni (2008) Proxy-based reconstructions of hemispheric and global surface temperature variations over the past two millennia. *Proceedings, National Academy of Science*, 105, 13252–13257.

Mann, M. E. (2012) *The Hockey Stick and the Climate Wars*. Columbia University Press: New York.

Martinez, O. (2016) *A History of Violence: Living and Dying in Central America*. Verso Books: London, New York.

Mayer, J. (2013) A word from our sponsor. *New Yorker Magazine*, May 27.

Mao, A., C. G. A. Harrison, and T. H. Dixon (1999) Noise in GPS coordinate time series. *Journal of Geophysical Research*, 104, 2797–2816.

Marcott, S. A., J. D. Shakun, P. U. Clark, and A. C. Mix (2013) A reconstruction of regional and global temperature for the past 11,300 years. *Science*, 339, 1198–1201, doi: 10.1126/science.1228026.

National Assessment Synthesis Team (2000) *Climate Change Impacts on the United States: The Potential Consequences of Climate Variability and Change*. US Global Change Research Program: Washington, DC.

Oreskes, N., and E. M. Conway (2010) *Merchants of Doubt: How a Handful of Scientists Obscured the Truth on Issues from Tobacco Smoke to Global Warming*. Bloomsbury Press: New York.

Piguet, E., and F. Laczko (eds.) (2014) People on the move in a changing climate. *Global Migration Issues*, 2, doi: 10.1007/978-94-007-6985-4.

Plass, G. N. (1959) Carbon dioxide and climate. *Scientific American*, 201, 1, 41–47.

Proulx, A. (2016) *Barkskins: A Novel*. Scribner: New York, London, Toronto.

Rahmstorf, S., and D. Coumou (2011) Increase of extreme events in a warming world. *Proceedings, National Academy of Science*, 108, 17, 905–917.

Raymo, M. E., and J. X. Mitrovica (2012) Collapse of polar ice sheets during the stage 11 interglacial. *Nature*, 483, 453, doi: 10.1038/nature10891.

Robbins, S. (2015) New wave of migrant children travelling alone: US flooded with children fleeing Central America. *Associated Press*, December 25.

Ruddiman, W. F. (2013) The Anthropocene. *Annual Review of Earth and Planetary Science*, 41, 45–68.

Schwartz, R. D. (2012) An astrophysicist looks at global warming. *GSA Today*, 22, 44–45.

Somerville, R. C. J. (2008) *The Forgiving Air: Understanding Environmental Change*, 2nd edn. AMS Books: Boston.

Somerville, R. C. J., and S. J. Hassol (2011) Communicating the science of climate change. *Physics Today*, 64, 48.

Vallis, G. K. (2012) *Climate and the Oceans*. Princeton University Press: Princeton, NJ.

Wasdin, H. E., and S. Templin (2011) *Seal Team Six: Memoirs of an Elite Navy Seal Sniper*. St Martin's Press: New York.

Wolff, M. (2008) *The Man Who Owns the News: Inside the Secret World of Rupert Murdoch*. Broadway/Random House: New York.

Yang, Q., T. H. Dixon, P. G. Myers, J. Bonin, D. Chambers, and M. R. van den Broeke (2016) Recent increases in Arctic freshwater flux: impacts on Labrador Sea Convection and Atlantic overturning circulation. *Nature Communications*, 7, 10525, doi: 10.1038/ncomms10525.

## CHAPTER 9

Dixon, T. H., Y. Jiang, R. Malservisi, R. McCaffrey, N. Voss, M. Protti, and V. Gonzalez (2014) Earthquake and tsunami forecasts: Relation of slow slip events to subsequent earthquake rupture. *Proceedings of the National Academy of Sciences*, 111, 48, 17039–17044.

Hofmeister, J. (2010) *Why We Hate the Oil Companies*. St Martin's Press/Palgrave Macmillan: New York.

Jiang, Y., T. H. Dixon, and S. Wdowinski (2010) Accelerating uplift in the North Atlantic region as an indicator of ice loss. *Nature Geoscience*, 3, 404–407, doi:10.1038/ngeo845.

Krabbenhoft, D. P., and E. M. Sunderland (2013) Global change and mercury. *Science*, 341, 1457.

Lubick, N. and D. Malakoff (2013) With pact's completion, the real work begins. *Science*, 341, 1443.

Newman AV (2011) Hidden depths. *Nature*, 474, 441–443.

Service, R. F. (2015) Tailpipe to tank. *Science*, 349, 1158–1160.

Streets, D. G., M. K. Devane, Z. Lu, T. C. Bond, E. M. Sunderland, and D. J. Jacob (2011) All time releases of Mercury to the atmosphere from human activities. *Environmental Science and Technology*, 45, 10485–10491

Toffler, B. L., and J. Reingold (2003) *Final Accounting: Ambition, Greed and the Fall of Arthur Andersen*. Currency/Doubleday/Random House: New York, London, Toronto.

# Index

*Note: An "A" in page number sequences refers to relevant content found in the book's online Appendix, available at www.cambridge.org/dixon.*